Appraising Sustainable Development

D1743198

Books in the Series

Integrated Water Resources Management in South and South-East Asia
Asit K. Biswas, Olli Varis, and Cecilia Tortajada (eds) (2005)

Water as a Focus for Regional Development
Asit K. Biswas, Olcay Ünver, and Cecilia Tortajada (eds) (2004)

Water Policies and Institutions in Latin America
Cecilia Tortajada, Benedito P.F. Braga, Asit K. Biswas and Luis E. García
(2003)

Integrated River Basin Management
The Latin American Experience
Asit K. Biswas and Cecilia Tortajada (2001)

Conducting Environmental Impact Assessment for Developing Countries
Asit K. Biswas and Modak Prasad (2001)

Sustainable Development of the Ganges-Brahmaputra-Meghna Basins
Asit K. Biswas and Uitto I. Juha (2001)

Women and Water Management
The Latin American Experience
Cecilia Tortajada (ed.) (2000)

Core and Periphery
A Comprehensive Approach to Middle Eastern Water
Asit K. Biswas, John Kolars, Masahiro Murakami, John Waterbury, and
Aaron Wolf (1997)

National Water Master Plans for Developing Countries
Asit K. Biswas, César Herrera-Toledo, Héctor Garduño-Velasco, and
Cecilia Tortajada-Quiroz (eds) (1997)

International Waters of the Middle East
From Euphrates-Tigris to Nile
Asit K. Biswas (ed.) (1994)

Appraising
Sustainable Development

Water Management
and
Environmental Challenges

edited by
Asit K. Biswas and Cecilia Tortajada
Third World Centre for Water Management
Atizapan, Mexico

BIBLIOTHECA ALEXANDRINA
مكتبة الإسكندرية

OXFORD
UNIVERSITY PRESS

OXFORD
UNIVERSITY PRESS

YMCA Library Building, Jai Singh Road, New Delhi 110001

Oxford University Press is a department of the University of Oxford. It furthers the
University's objective of excellence in research, scholarship, and education
by publishing worldwide in

Oxford   New York
Auckland   Cape Town   Dar es Salaam   Hong Kong   Karachi
Kuala Lumpur   Madrid   Melbourne   Mexico City   Nairobi
New Delhi   Shanghai   Taipei   Toronto

With offices in
Argentina   Austria   Brazil   Chile   Czech Republic   France   Greece
Guatemala   Hungary   Italy   Japan   Poland   Portugal   Singapore
South Korea   Switzerland   Thailand   Turkey   Ukraine   Vietnam

Oxford is a registered trade mark of Oxford University Press
in the UK and in certain other countries

Published in India
by Oxford University Press, New Delhi

© Oxford University Press 2005

The moral rights of the author have been asserted
Database right Oxford University Press (maker)

First published 2005

ISBN-13: 978-0-19-566891-9
ISBN-10: 0-19-566891-X

Typeset in AGaramond 11/13.6 by InoSoft Systems, Noida
Printed by Pauls Press, New Delhi 110 020
Published by Manzar Khan, Oxford University Press
YMCA Library Building, Jai Singh Road, New Delhi 110 001

Contents

Preface

The book considers two challenging issues of the twenty-first century, the implementation potential of the concept of sustainable development, and its possible application to make water management more efficient and equitable. Both are complex issues, which have not been objectively and comprehensively examined thus far.

There is no question that the concept of sustainable development has been popular for nearly a quarter of a century. In fact, sustainable development has been so popular that to ask to what extent the application of this concept has improved resource management practices during the past two decades, or improved the quality of life of the poor people in the developing world, or improved environmental conditions, are now considered in many quarters to be equivalent of lese-majesty! In fact, many international organizations and most major foundations have programmes on some aspects of sustainable development. Inspite of this popularity, however, one fundamental question which has never been asked, let alone answered, is if this popular concept has made any difference to the world, or if the world would be any different if this paradigm had not been in use?

The book attempts to answer the above question and many others in terms of water management. It is now widely believed that water is likely to be one of the most critical resource issues of the coming decade, or decades, both in terms of quantity and quality. Under the present management practices, many regions of the world are facing water scarcities, and most international institutions like the various United Nations agencies and the

World Bank, have claimed that these scarcities will escalate in the future, creating serious problems for the humankind as well as the environment. While many pronouncements have been made on the impending global water crisis, most of these have been made on the basis of forecasts based on unavailable and/or unreliable data, assumptions of primarily incremental changes in the current business-as-usual practices (inspite of the rhetoric that claims otherwise), incomplete and faulty analyses, and the absence of a realistic and forward-looking vision. Based on a comprehensive analysis of the current and the future trends, it can be said with considerable certainty that within the next two decades the following developments are likely to occur, which will influence the management practices in the water sector, through numerous direct and indirect pathways, some of which are predictable but others are not.

- The world of water management is likely to change more during the next 20 years compared to the past 2000 years;
- Many of these changes are likely to occur due to forces that are mostly outside the control of the water sector;
- While water scarcity will remain a problem in some parts of the world, it is unlikely to be of crisis proportions for the world as a whole;
- Water crisis is more likely to occur because of deterioration of its quality in nearly all over the developing world (for example, currently less than 10 per cent of contaminated water is properly treated in most developing countries), and non-availability of investment funds in a timely manner in developed and developing countries to manage all water quantity and quality issues properly. While the water profession and the media have focused almost exclusively on scarcity as the reason for a water crisis, real crisis is likely to be triggered by continuing water quality deterioration and lack of investment funds, none of which is receiving adequate attention at present.
- The complexities associated with water management will increase very significantly in the future.

- The potential impacts of technological developments in the water and related sectors, like biotechnology and desalination, have been mostly ignored at present, as well as the implications of globalization, communication and information revolution, and the use of different economic instruments for managing water.

It is safe to predict that the water problems of the future will become more and more interlinked with other development-related sectors like agriculture, energy, industry, transportation, and communication, and social sectors like education, environment, and health. There is no question that water policies and major water-related issues of the future should be assessed, analysed, and resolved within an overall societal and development context. Otherwise, the main objectives of water management, such as improved standard and quality of life of the people, poverty alleviation, regional income redistribution, and environmental conservation cannot be achieved.

While the challenges facing water management can be predicted with a fair degree of accuracy, a major unknown at present is how can these challenges be successfully met in a socially-acceptable, economically-efficient, and environmentally-friendly manner. A popular answer has been that these challenges can be met by applying the concept of sustainable development to the water sector. The book objectively analyses how realistic is this expectation, without any dogmas, preconceived ideas or hidden agendas.

What is sustainable development? There is no agreement at present as to what this really means. Dozens of definitions exist at present, but one that is now extensively used is by the so-called Brundtland Commission, which defined it as 'development that meets the needs of the present without compromising the ability of the future generations to meet their own needs'.

This definition, at least on a first reading, appears to be reasonable, equitable, and well-meaning. However, the question that arises is if this well-intentioned and feel-good definition has any real meaning in terms of its application and implementation to improve existing water management practices, or if it is just a collection of good-sounding and trend words which collectively provide a somewhat amorphous definition which is of limited, or even of no help, to plan and manage water resources projects and programmes.

Objective and in-depth analyses indicate that, for all practical purposes, the definition formulated by the Brundtland Commission cannot be implemented, not just in the water sector, but also any other development sector. Not surprisingly, even though the rhetoric of sustainable water resources development has been very strong in numerous national and international forums during the past two decades, its actual use in the real world, irrespective of what it means, has been minimal, even indiscernible in most cases. In fact, based on evidences available at present, it can be argued that even if the paradigm of sustainable development did not exist, the world of water management probably would have looked very similar to what it is at present.

The concept of sustainable water development is now widely considered essential in most countries of the world, including international institutions. As noted and elaborated in this book, this is despite the fact that it is not operationally possible to plan, implement and manage a water resources system in such a way that it becomes inherently sustainable, however this may be defined right from the very beginning. Even after such widespread endorsement of the paradigm, it is still not possible to identify what are the parameters that should be measured, which will indicate a water resources system is sustainable, non-sustainable, or in transition between sustainability and non-sustainability.

The universal popularity of a vague concept that defies definition and implementation is not new in the area of resources management. For instance, during the twentieth century, many such concepts have come and gone, without leaving much of a footprint as to how resources could be efficiently managed on a long-term basis. In fact, it can even be argued that the vagueness of a concept to a significant extent increases its popularity, since people can continue what they have been doing before, but at the same time claim that they are following the latest paradigm. The fact that it has not been possible to operationalize this concept in over twenty years raises another fundamental question: is it an universal solution as its numerous proponents currently claim it to be, or is it a concept that has limited implementation potential, irrespective of its conceptual attractiveness and current popularity? Unless the concept of sustainable water resources management can actually be applied in the real world to demonstrably improve the existing water management practices, its current popularity and extensive endorsements by international institutions become irrelevant. Knowledge, fortunately, does not advance by consensus or popularity: if it did, we would still be living in the Dark Ages!

In addition, the world is heterogeneous, with different cultures, social norms, physical attributes, climatic conditions, a skewed availability of renewable and non-renewable resources, investment funds, management capacities, and institutional arrangements. The systems of governance, legal frameworks, decision-making processes, and types and effectiveness of institutions often differ from one country to another in very significant ways. Accordingly, and under such diverse conditions, one fundamental question that needs to be asked is that if it is possible for a single paradigm of sustainable water resources management to encompass all countries, or even regions, with diverse physical, economic, social, cultural, and legal conditions? Can a single paradigm of sustainable water resources management be equally valid for an economic giant like the United States, technological powerhouse like Japan, and for countries with as diverse conditions as Bhutan, Brazil or Burkina Faso? Can a single concept be equally applicable for Asian values, African traditions, Japanese culture, Western civilization, Islamic customs, and emerging economies of the Eastern Europe? Can any general paradigm be equally valid for monsoon and non-monsoon countries, deserts and very humid regions, and countries in tropical, sub-tropical, and temperate regions, with very different climate, institutional, legal, and environmental regimes? The answer most probably is likely to be no.

In this book, some of the world's leading intellectuals and experts analyse the current status of application of the sustainable development paradigm to the water sector. The authors come from different backgrounds, expertise, disciplines and institutions from different parts of the world. Their overwhelming conclusion is that irrespective of the current popularity of the sustainable development concept, its application in terms of improving efficiency of water management has been minimal. This status is unlikely to change in the foreseeable future.

I would like to take this opportunity to pay tribute to the Sasakawa Peace Foundation of Japan, without whose support this analysis simply would have not been possible. When the Third World Centre for Water Management first decided to evaluate objectively and comprehensively the current status of application of sustainable development to the water sector, we were actively discouraged by most institutions. The head of a major international institution tried to dissuade the Centre from doing this study. His thesis was that the concept is so popular that it would not be desirable for our Centre to question it. His view was that 'even if you prove that its current status of application leaves much to be desired, the Centre will be the looser since some funding agencies may decide not to support its projects in the future'. The head of a major international foundation told me that for several years their institution has had a programme on sustainable development. He could not put forward any proposal which may question the concept, since his trustees may ask some very awkward questions, including the question 'if the emperor had any clothes'.

We thus applaud Sasakawa Peace Foundation for their support to this study, which most other institutions did not want to consider since the results could question the appropriateness and relevance of a sacred cow. We are especially thankful to Mr Akira Iriyama, President of the Sasakawa Peace Foundation, who inspite of his numerous commitments, participated in the entire workshop which assessed the concept. We are also grateful to Dr Takashi Shirasu, who was the Senior Programme Officer when we started to discuss the project with the Foundation, for his intellectual support, and to Dr Mihaela Serbulea, who not only participated and contributed to the workshop but also was the programme officer responsible for the implementation of the project. This book simply could not have been produced without the intellectual support of Mr Iriyama, Dr Shirasu, and Dr Serbulea.

The workshop to review the papers was organized at the Alexandria Library in Egypt. We very much appreciate the support given to us by Dr Ismail Serageldin, President of the Library, and Dr Mahmoud Abu-Zeid, Minister of Water Resources and Irrigation of the Government of Egypt, to organize the workshop in Alexandria. Thanks to their support and outstanding hospitality, the workshop became a memorable event for all the participants.

Last, but not least, I would like to thank Dr Cecilia Tortajada and Ms Thania Gomez of our Centre for all the internal work which made the project so successful and productive. The excellent work done by my daughter, Andrea Lucia, for the proof-reading and indexing of this book is much appreciated. She certainly did a much better job than I would have ever done!

Asit K. Biswas, President
Third World Centre for Water Management
Atizapan, Mexico

Foreword

On behalf of the Sasakawa Peace Foundation, I am pleased to write the Foreword to this important book, which aims to reassess the applicability of the sustainable development paradigm to the water sector.

Being a layman on the water issue per se, I do not even pretend to know the current priority problems or non-problems of the water sector. For instance, I cannot tell which of the following arguments is valid. Some argue that water scarcity is the issue whereas others argue that the issue is not its scarcity itself but its mismanagement. Some documents state that more people are drinking clean water as compared to the number 20 years ago whereas others refute this claim. Also, we hear an argument that the issue lies not in water itself but in the order of priority to be put on water among other issues like environment, poverty alleviation, and human rights. Even within the water sector itself, a number of zero-sum kind of trade-offs between regions, countries, or social sectors within a country are witnessed. It will not be productive to continue with a list of these anecdotal examples since, supposedly, answers to these questions are all too obvious to the water specialists. Instead, I will briefly touch upon an overarching paradigm over these questions that is the core subject of this book, namely sustainable development.

As in the case of other buzzwords such as 'social capital', 'civil society', and 'human security', the term 'sustainable development' has been used for many years with varying definitions. For that matter, this paradigm has been used and valued because it can have a number of interpretations according to various players who may have their own vested interests.

Let me elaborate this point a little further. When the idea of sustainability was first introduced to the field of fisheries much earlier, it appeared to be rather simple and practical, since it meant that the amount of catch should be equal to or less than its total reproduction rate. In comparison, when it was later widened and broadened by the so-called Brundtland Commission as 'the development that meets the need of the present without compromising the ability of the future generations to meet their own needs', it no more was meant to be a practical, operational definition. Rather, the words 'sustainable development' became the statement for the desirable future for mankind. It is natural then to expect this statement to mean 'all things to all men', which by itself is difficult to deny, and which will keep being an eternal goal for all. However, at the same time, we all know it is not too practical to expect this to be achieved today. Are there then any possibilities for the concept of sustainability to become practical and operational again?

I am reminded here of the words of Dr Norman Borlaug, the esteemed Nobel laureate agronomist, who once told me, 'sustainable development' means nothing unless three things are defined. These are: for how many people, with what living standards, and within what time period. The implication of his words seemed to me to be 'We cannot wait forever to achieve all mankind to enjoy optimum living standards. Instead, now is the time to agree upon a practical and operational definition of sustainable development. We cannot leave this paradigm to stay as an empty political slogan.'

I was pleased, therefore, to find the following lines in Prof. Biswas's paper (Chapter 3, 'Sustainable Development: Some Unanswered Questions'): 'In spite of the widespread use of the paradigm of sustainable water development, it has to be admitted that even after some 25 years of use, it has not been possible to define any water development process which could be planned and implemented in such a way from the very beginning so that it could become inherently sustainable, however it may be defined. Nor is it possible to identify all the appropriate parameters that should be monitored and evaluated to indicate the beginning of a transition process from sustainability to unsustainability, and vice versa. After some 20 years of practice, it is still not known how sustainability can be measured, analysed, judged, or implemented, in the context of water development. Such critical issues are not even raised at present, let alone discussed and solved.' If we were to discuss sustainable development in line with his discourse, no doubt the outcome will be very beneficial not only to water specialists, but also to all those, including myself, who are engaged in development at large.

In pursuing this line of undertaking, I sense two elements that are crucial to the process. Again, I am referring to them not from the water-specific point of view solely, but rather from a general developmental perspective.

The first element is cultural diversity, or the differences in the ways and manners in which people place importance on various issues, that is, there is no single panacea that can treat all symptoms. There tends to exist, however, an innocent optimism that one successful *modus operandi* can and should work elsewhere, too. In some cases it extends even further to the belief in the single normative way of development. This may sound absurd, but we have witnessed many developmental experiments that have advocated one single prescription to a number of countries regardless of their differences in the system of governance, legal frameworks, decision-making processes, effectiveness of management among others. In the water sector, it is my hope that the water experts have not experienced too many such cases.

This element leads me to the second one. Let me go back to Norman Borlaug's comments mentioned earlier. For whom, with what level, and when is sustainable development to be realized? The well-being of future generations is important. We have to avoid deprivation of the wealth of our children, or grandchildren, or great-grandchildren. But it should not mean a total refusal to meet the pressing needs of our time. When we raise, or question, this issue in either–or format, we are already in the realm of political, not scientific, arguments. The need for an objective, non-dogmatic scrutiny of the developmental paradigm has been never stronger than today.

Without doubt, this exercise of reassessing the sustainable development paradigm is fruitful because, contrary to what many may believe, the articulation of and debate on paradigms and ideologies may be the quickest way to improve practical methods of development. And, in doing so, we will be eventually providing an answer to the three questions raised by Dr Norman Borlaug as mentioned above.

The Sasakawa Peace Foundation considers objective, critical, and comprehensive discussions of the priority development paradigms of the present, an important consideration for the future. The conceptual attractiveness of a paradigm is not enough: it must be applicable in the real world to improve the well-being of both the present and the future generations. We are thus pleased to support the activities that have led to the publication of this book. I am confident that the book will act as a creative eye-opener to many people.

Akira Iriyama, President
Sasakawa Peace Foundation
Tokyo, Japan

Table and Figures

Table

Figures

1

Sustainable Development: A Critical Assessment of Past and Present Views

Cecilia Tortajada

Introduction

As long back as more than three decades, the need for a different type of economic growth that is more efficient in terms of use of non-renewable resources and less harmful to the physical environment had been noted. It has been argued that such a new type of development process is necessary because of the limited availability of natural resources, as well as the limited absorptive capacities of the ecosystem to assimilate waste products (OECD 1979).

During the 1970s, major international organizations such as the United Nations Environment Programme (UNEP), expressed their concerns about the need for a new kind of development where the implications for rich and poor countries are recognized, which presupposes new directions for growth and development, nor their cessation, and which incorporates the environmental dimension, approached accordingly by industrialized and developing countries. However, since both the objectives of environmental and development policies are to improve the quality of life of the populations, environment should play a central role in development policies. It is essential to relate development to the opportunities and limitations created by the natural resource base to all human activities. New patterns of development are necessary, because the previous or actual ones have resulted in environmental degradation, because of increasing social inequalities and because they have not met the expectations of the people in the developing world. Of course, necessary changes are immense and would require years to carry through. However, it is necessary to start (Tolba 1982).

There has thus been a long-term concern on the concepts of development, and less clearly, on the implementation and impacts of these concepts. For more than three decades, it has been argued that environment-related issues should be integral components of planning and policy-making because of their impacts on the quality of life of the populations. Such environmental concerns are not new: they have been around for decades, or even centuries, in one form or another. The main issue thus is not whether such concerns have existed, but rather to what extent they have been mainstream views, and whether these environmental issues have been taken into consideration in the planning and policy-making process of various countries. An objective assessment would indicate that there have been extensive and intensive discussions on the importance and relevance of environmental issues that are related to the development process. However, what has been actually achieved lags far behind the international rhetoric.

During these decades, the focus of discussions has changed, terminology and concepts have evolved, discourses on specific issues have ebbed and flowed, and new and modified paradigms have been proposed. However, development practices have had limited impacts on poverty alleviation, and the situation from an overall environmental viewpoint has been worsening continuously. In other words, national and international leaders and their institutions have not followed up their words with matching deeds.

One such example is the United Nations Conference on Environment and Development (UNCED) organized in Rio de Janeiro in 1992, where virtually all the leaders of the world supported the principle of sustainable development. Later on, in

1997, the implementation of the Action Plan approved by all the governments at Rio, known as Agenda 21, was assessed at a Special Session on Sustainable Development of the United Nations General Assembly. According to this Special Session, implementation of the sustainable development concept required political commitments, from which the leaders have 'shied away'. This session was useful in the sense that 'it brought home the uncomfortable truth that sustainability requires changes to deeply rooted modes of political behaviour' (Jordan and Voisey 1998).

An important question then is how should the long-term development plans for countries and their populations be formulated and implemented, within which environmental aspects can be seriously considered? In other words, when if ever, will it be realized that the environmental aspects are intertwined with economic and social issues, which will certainly impact on the humans? (Hammond 1998)

Accordingly, it is necessary that the environmental thinking evolves and that it goes hand in hand with development policies. The environment needs to be recognized as an important factor which would assure the sustainability of the development processes themselves. However, this can only be realized if the environmental considerations go beyond discourses and statements, and become an integral component of the development process itself.

In terms of water policy formulation and implementation, most developing countries face fundamental problems that relate to issues as basic as the definition of goals and objectives (ECLAC 1998). Both the planning and management of natural resources, water included, are plagued and constrained by concepts that often cannot be implemented because they cannot be properly defined and thus operationalized. In spite of these shortcomings, governments often insist on paying homage to certain paradigms, irrespective of their implementation potential, simply because they are part of the current global thinking. For example, available evidences indicate that sustainability represents more of a concept than an implementable reality (Dragun and Jakobsson 1997; Meppem and Bourke 1999; Meppem and Gill 1998). Hence, it is somewhat unlikely that any government pursuing sustainable development, as it is defined at present, would be able to develop realistic plans that can be properly implemented.

The urgent need to move from concept to implementation of any paradigm is of utmost importance to the water professionals. Global paradigms like sustainable development and integrated water resources management are unquestionably conceptually attractive but their actual implementation in operational terms has left much to be desired. It is thus essential to objectively analyse their applicability: conceptual attractiveness alone is not a solution. Rather than ignoring the need for alternative conceptual frameworks which are implementable, individuals and institutions collectively should welcome constructive analyses and criticisms of the existing mainstream approaches. Some of the current conceptual frameworks and theories on water development should thus be carefully analysed and, if necessary, reconsidered. Such analyses and open discussions can only be beneficial to the water profession, resulting also in more efficient water management.

The objective of this chapter is to analyse the effectiveness of some global paradigms in the field of water, as well as the possibility of moving from concept to implementation, in terms of improving water management processes and practices. Many concepts are used extensively at present, for example sustainable development, environmental sustainability, integrated water resources management, or integrated river basin management. However, this chapter will focus only on the concept of sustainable development.

In spite of the current popularity and widespread mention of the concept of 'sustainable development', its origin is not well known. Thus, a brief review of its origin and its evolution is presented. The analyses show that even though many developing countries have adopted the global views in theory, they still need to strengthen their institutions, implement legislations, develop long-term policies, and build administrative, technical, and management capacities so that at least, significant parts of the theories can actually be translated into effective practices.

Sustainable Development

Evolution of the Concept

Even though the concept of sustainability has been used extensively since the mid-1980s, the idea is not new. For example, the term 'sustainability' has been widely used in fisheries and forestry for nearly a century to define long-term management techniques for harvesting reproducible natural resources. Thus, terminologies like maximum safe yield have been common for many decades in the fields of fisheries and forestry.

Contrary to the popular view, the concept of sustainable development did not start with the publication of the report of the World Commission on Environment and Development (WCED, or the Brundtland Commission Report) in 1987. In fact, by the mid-1980s, well before this report was published, the concept of sustainable development had already become popular, initially through the work of the UNEP, and later by the activities of the World Bank.

The earliest reference to the concept of sustainable development, as well as the use of this terminology, goes back to at least over half a century. It is possible that other authors may have used this terminology before 1948, even though no such reference was found during the course of research for this chapter.

In 1948, Fairfield Osborne, the founder and the then President of the Conservation Foundation, wrote in his book *Our Plundered Planet*:

We are rushing forward unthinkingly through days of incredible accomplishment . . . and we have forgotten the earth, forgotten it in the sense that we are failing to regard it as the source of our life.

Osborne was concerned with the 'accumulated velocity with which (man) is destroying his own life sources.' He insisted that the only kind of development that makes sense is 'development that can be sustained'.

Intellectually, however, the concept of sustainable development was promoted by the UNEP, which was established in Nairobi, Kenya, as a direct result of the United Nations Conference on the Human Environment that was held in Stockholm, Sweden, in June 1972. A small group of environmental scientists, meeting in Nairobi in 1975, under the aegis of UNEP, extended the concept of sustainability from fisheries and forestry to the development process itself.

Shortly after this meeting, in 1976, Mostafa Kamal Tolba, the then Executive Director of UNEP, in an address in London pointed out (Tolba 1982):

A new kind of development is needed because it is essential to relate development to the limitations and opportunities created by the natural resource base to all human activities. It is also required because it is now clear that past patterns of development in both developed and developing countries have been characterized by such serious environmental damage that they are simply not sustainable.

Tolba (1982) then went on to argue:

The most pressing objective of environmental management is to meet basic human needs within the potentials and constraints of environmental systems, including natural resources. Environmental management brings two new dimensions to the development process: it broadens the concept to include environmental quality, and it expands it in time to include development over the long-term on a sustainable basis.

Tolba's eloquent arguments for a new form of development process that is sustainable over the long term touched a chord in the environment movement. In 1981, A. W. Clausen, the then President of the World Bank, gave a major statement on 'Sustainable Development: the Global Imperative' (Clausen 1981). A year later, during the commemoration of the tenth anniversary of the Stockholm Conference, in Nairobi on 10–18

May 1982, the world community of states unanimously recommended 'sustainable socio-economic development'. The Nairobi Declaration, that resulted from the commemorative meeting, concluded by urging (Tolba 1988):

... all Governments and peoples of the world to discharge their historical responsibility, collectively and individually, to ensure that our small planet is passed over to future generations in a condition which guarantees a life in human dignity for all.

In 1987, in its report entitled 'Our Common Future', the WCED recommended the concept of sustainable development, which it loosely defined as 'development that meets the needs of the present without compromising the ability of the future generations to meet their own needs'.

Even though the WCED report made continual references to sustainable development, it was totally silent on how the concept could be operationalized. Sustainable development was expected to be achieved in an unspecified and undetermined way, some time in the future. Nor did the definition include the realization of an equitably distributed level of economic well-being, without which no development can be sustained over the long term. The issue of equity is especially important for developing countries (Biswas 1997).

The United Nations General Assembly considered both the WCED report (1987) and a report on 'Environmental Perspective to the Year 2000 and Beyond' prepared by UNEP (UNGA 1987). In the General Assembly Resolution 42/186 of 1987, it noted that 'different views exist on some aspects' between the WCED and UNEP reports. It, however, welcomed:

... as the overall aspirational goal for the world community the achievement of sustainable development on the basis of prudent management of available global resources and environmental capacities and the rehabilitation of the environment previously subjected to degradation and misuse....

Following the work of the UNEP and the WCED, and the passing of the above-mentioned United Nations General Assembly Resolution, sustainable development became 'the' paradigm for development. The various United Nations agencies, all the development banks and the bilateral aid agencies, and nearly all the governments, embraced the paradigm of sustainable development, even though it was never properly defined, except in a broad and general way. Additionally, no serious discussion ever took place as to how the concept could be operationalized in the real world, so that a development process could be planned and managed from the very beginning in order for it to become inherently sustainable.

The ideological debate about ways of integrating environmental considerations into policy-making also did not start with the publication of the UNEP and the Brundtland Commission reports. During the 1970s and 1980s, attempts were made to articulate alternatives to an almost exclusive reliance on conventional indicators such as economic growth in terms of gross national product (GNP), balance of payments, employment, index of inflation, etc. Among other catchwords, 'qualitative growth' was one of the first to signal a new direction of societal interest. It was argued that growth exclusively in terms of GNP for some activities was incompatible with environmental goals, while growth in other activities (with related goods and services) was basically beneficial (Söderbaum 1998).

The Discourse

As mentioned before, in response to the perceived threat of impending ecological crisis during the post-1970 period, a dominant environmental discourse *constructed* itself. Certain words were favoured for evoking images of consensus, unity, and common purpose, like sustainability, diversity, democracy, community, globalization, and environment (Bourke and Meppem 2000). As a consequence, the concept of sustainable development, as well as the mechanisms to address it, became very important issues. So far, however, there is still no agreement even on the meaning or definition of sustainable development. Thus, it is not surprising that little consensus exists with regard to formulating and operationalizing sustainable development policies, except in broad and general terms (Biswas 1996; Dragun and Jakobsson 1997; Goodland 1997; Meppem 2000; Meppem and Bourke 1999).

The ongoing debate about sustainable development and its various meanings is considered to be very much ideological (Söderbaum 2000). The diversity of discourses on sustainable development may not necessarily reflect conflicts over content, but on interests and in opinions on the processes through which the different sectors of the societies want to assure that their own needs and interests are represented in the development decision-making. Thus, sustainable development may not refer to a quantifiable goal that can be achieved at any specific moment in time: it may refer instead to the possibility of establishing a balance between environmental, social, and economic interactions. This process, at least in theory, is expected to improve the quality of life of human beings, and simultaneously maintain the integrity of the environment.

Sustainable development may be considered to be more of a desirability with regard to future human development, in which case it may represent a constraint to the present development. Sustainable development may assure certain life opportunities in the future, but at the cost of the modification or sacrifice of life opportunities in the present. The concept of sustainable development at the first instance may appear somewhat simple, but in reality it is very complex. This is because it is expected to result from a series of decisions taken by several generations of human beings in different parts of the world, at different levels of governments, with changing socio-economic conditions, differing cultural values, uncertainties, and socio-economic goals which are seldom shared by all the members of the different societies, since people tend to work at the individual level (Dourojeanni 1999). In addition, the nation-states have their own interests, which may vary with time. This complexity may result in a permanent gap between the current understanding and the one necessary to address evolving economic, social, and environmental planning and management issues comprehensively, as well as the institutional, legal, and even participatory considerations. Working with the concept of sustainable development means embracing ambiguity, since it deals with societal values, which are diverse, and may often vary with time. If the conflicts in interpretations of sustainable development reflect the diversity of the concerns and interests of the populations in time and space, it is fundamental then to learn how to accommodate the politics of these divergent claims for attention. Additionally, if the relations between citizens and the private and public sectors are increasingly interdependent, necessary processes and polices should be developed in order to approach the various interests from an integral viewpoint (Meppem 2000; Meppem and Gill 1998).

Bottlenecks for its Implementation

To design appropriate policies for sustainable development, the goals must be expressed in terms of specific indicators. However, these choices are to a certain degree subjective by nature, and are dependent to a great extent on the cultural preferences and interests of an individual, a community, or a country. This implies that different societies, with differing social, economic, and cultural conditions, may choose different sustainability criteria and may even select different paths to sustainability (Raskin et al. 1998). Thus, one of the greatest difficulties in achieving sustainable development lies in the lack of indicators for its measurement, since none of the three objectives of sustainable development (economic, environmental, and social) is currently measured using compatible parameters. The indicators used to quantify the economic, social, and environmental objectives do not have a common denominator, nor do universal conversion formulae exist: economic growth is measured using economic indicators, social equity is determined on the basis of social parameters, and environmental protection is measured on physical and biological terms. Given the absence of suitable indicators, and the fact that each of these objectives is measured according to different criteria, it does not seem that it would be possible to interlink the three objectives in a single plane. Quantification of economic, social, and environmental objectives may not be possible, unless compatible quantifiable parameters are available for all the three sectors (Dourojeanni 1997).

At the same time, sustainable development would not be achieved if emphasis was placed on either of the economic, social, or environmental objectives at the expense of the others. Thus, the stakeholders must contribute simultaneously to economic growth, social equity, and environmental protection, most likely through trade-offs, negotiations, and by modifying everyday practices. The agreements between the various stakeholders are likely to be more productive, equitable, and workable if there is an understanding of the actual value of the specific resources and products for each one of them (Dourojeanni 1997). However, values are often subjective, and hence inter-comparison of subjective values can be a most difficult task under the best of circumstances.

Further issues confronting sustainable development are the risks and uncertainties that are inherently associated with complex systems. For example, it is now universally accepted that food production must be maximized to feed an expanding population base in developing countries. Accordingly, resources such as land and water must be used intensively to maximize food production. Hence, a fundamental question, for which there is no clear-cut answer at present, is up to what level can the food production system be intensified, in terms of land and water, without sacrificing sustainability? There are other difficult questions as well. For example, in the area of water, what early warnings could indicate the beginning of a transition process from sustainability to unsustainability? What parameters should be monitored to indicate that such a transition is about to occur, or indeed is occurring? Existing knowledge bases and databases are inadequate even to identify all the relevant parameters that could indicate passage from one stage to another. Thus, concurrently it is not possible to accurately detect, much less predict, the transition of a sustainable system to an unsustainable one and vice versa (Biswas 1996).

In order to formulate and implement sustainable water development policies, the developing countries require much more knowledge, expertise, data, and information than they currently possess. Thus, one of the first priorities should be to broaden their knowledge and information bases in the technical, economic, social, and environmental fields. Research, training, and capacity building, both for individuals and institutions, should be developed, keeping in mind the type of environmental problems that are likely to be faced during the process of water development over the course of the next several decades. Developing nations should base their development agendas on their own administrative, technical, scientific, and economic capacities. For water development to be more effective, disciplines should be approached from a broader perspective, and knowledge should be developed in such a way that it will be useful to decision makers outside the academic and research fields (Serageldin et al. 1998).

In terms of technology, it is important to remember that while it may have a major impact on the global development process, it may not necessarily solve demographic, social, and environmental problems. The impacts of technology often depend upon its social context, in terms of whether, when, and how it is used. Technological innovations may have important economic effects like lowering costs through improved efficiency, making alternatives possible which were not feasible before, and accelerating economic growth. However, the development of new technology is often less important than its appropriate use. Whether technology will solve all, or most, water-related problems remains to be seen, since social factors have the definitive say in its implementation, and it may take decades for new technology to be adopted, and for societies to benefit from it (Hammond 1998).

The integration of the environmental concerns in development planning would require specific actions at the national levels. Some of the major policy areas may include location (or relocation) of industries, land use policies, community development, etc. Proper planning of infrastructure is important so that individual development projects are integrated within an overall framework for regional development planning and management. The social benefits and costs of projects, including their favourable and unfavourable impacts on the environment and the populations, should be fully reflected in these policies. Too often the negative impacts of many projects have been ignored in the initial planning stage, and so the awareness of the society of many of the environmental disruptions resulting from these projects has come at a very late stage, when the construction has already been completed, and the adverse impacts have already surfaced. Cost-effective alternatives available at such late stages to take ameliorative measures are likely to be limited. Accordingly, it is important to analyse comprehensively both the favourable and the unfavourable social and environmental impacts before implementing development projects, so that the society may be able to compare them against the economic and social benefits expected from the project. Feasible alternatives can then be considered (Modak and Biswas 2001; Tortajada 2000).

Concluding Remarks

Sustainable development has been a powerful and all-embracing slogan during the past 15 years, mainly in the international political fora. Every government is for it, as are all the major international organizations and non-governmental organizations (NGOs) working on issues related to the environment. This is so in spite of the fact that there is no agreement as to what is meant by sustainable development, whether it works, and if so, under what conditions, as well as what are its impacts (positive, negative, or neutral) on human lives, development indicators, and the environment. More important, this is true even though the concept of sustainable development, even several decades later, does not seem to have reached the policy-making level or have reduced the deterioration of the environment.

In addition, the world is heterogeneous, with different cultures, social norms, physical attributes, skewed availability of renewable and non-renewable resources, investment funds, management capacities, and institutional arrangements. The systems of governance, legal frameworks, decision-making processes, and types and effectiveness of institutions differ from one country to another in very significant ways. Furthermore, countries are at different stages of development, and thus their needs and requirements, which vary with time are also different. Accordingly, and under such diverse conditions, another fundamental question that needs to be asked is if it is possible that a single paradigm, that of sustainable development, can encompass all countries, or even regions, with diverse physical, economic, social, and cultural conditions. Is it feasible that a single paradigm like sustainable development be equally valid for technological giants like the United States and Japan, the world's most populous countries like China and India, and for countries as diverse as Burkina Faso and Mexico? Is it possible for a single concept to be equally applicable for Asian values, African traditions, Japanese culture, and Western civilization?

The point of departure for the development process is different from one country to another for technical, economic, historical, cultural, and other associated reasons. Regarding water resources, it is clear that each country needs to formulate its own water development strategies based on its specific conditions, requirements, and expectations. However, in many parts of the world, practices, processes, and legislations are being copied from other countries, without adapting them specifically to their own conditions. Institutional frameworks are being structured often according to the latest international thinking, without any detailed review of their applicability and usefulness within the national context.

In terms of environmental sustainability, irrespective of the rhetoric, and although most developing countries have tried to protect their image at the international level, and declare themselves to be for 'the environmental sustainability of water resources', poor management of water resources continues, and will continue, to have serious social, economic, and environmental implications at the local and national levels in both the short- and long-term future. Many times, such mismanagement has contributed to increasing poverty, and deterioration of the quality of life of the populations, especially in terms of health.

Many developing countries have claimed that the main constraint for fulfilling their commitments of Agenda 21 has been primarily lack of financial support. However, while lack of funds is certainly a problem, it seems that a greater limitation results from the absence of leadership, managerial and technical capacities, an almost exclusive top-down centralized approach, absence of stakeholder participation, and lack of any long-term vision in most fields, including water. Not surprisingly, progress in improving water management practices has been somewhat limited during the last 30 years in the developing world. In fact, much more could have been accomplished with the budgets that were available if the leadership had a clear vision as to what should be accomplished and their relative priorities. Not surprisingly, water problems of most developing countries have become more acute, especially in terms of water pollution.

Hence, regardless of the widespread use of the rhetoric of sustainable development and environmental sustainability, it has to be admitted that even after years of use, it has not been possible to define a development process which could be planned and implemented in such a way from the very beginning so that it could become inherently sustainable, however it may be defined. Nor has it been possible to identify the parameters that should be monitored and evaluated to indicate the beginning of a transition process from sustainability to unsustainability, and vice versa. After over 15 years of rhetoric, it is still not known how sustainability can be measured, analysed, judged, or implemented.

Any development expert knows, at least intuitively, that no single pattern of development is the most appropriate for all countries of the world at any specific point in history. There is no one single path to development, which can be successfully followed by all countries at all times. Thus, the fundamental question that needs to be asked, and unambiguously answered, is whether it is possible that one, and only one, single paradigm, that of sustainable development, is valid for the entire world.

References

Biswas, A. K., 1996, 'Water Development and Environment', in A. K. Biswas (ed.), *Water Resources, Environmental Planning, Management and Development*, McGraw Hill, New York.

———, 1997, 'Sustainable Water Development from the Perspective of the South: Issues and Constraints', in M. Abu-Zeid and A. K. Biswas (eds), *River Basin Planning and Management*, Oxford University Press, New Delhi.

Bourke, S., and T. Meppem, 2000, 'Privileged Narratives and Fictions of Consent in Environmental Discourse', *Local Environment*, vol. 5, no. 3, pp. 299–310.

Clausen, A. W., 1981, 'Sustainable Development: The Global Imperative', *Mazingira*, vol. 5, no. 4, pp. 2–13.

Dourojeanni, A., 1997, *Management Procedures for Sustainable Development (Applicable to Municipalities, Micro-regions and River Basins)*, Economic Commission for Latin America and the Caribbean, United Nations, Santiago.

———, 1999, *La Dinámica del Desarrollo Sustentable y Sostenible*, Comisión Económica para América Latina y el Caribe, Naciones Unidas, Santiago.

Dragun, A., and K. Jakobsson, 1997, 'Introduction, New Environmental Policy Dimension', in A. K. Dragun and K. M. Jakobsson (eds), *Sustainability and Global Environmental Policy, New Perspectives*, Swedish University of Agricultural Sciences, Edward Elgar, Cheltenham.

ECLAC, 1998, *Reflections on Territorial Strategies for Sustainable Development*, Economic Commission for Latin America and the Caribbean, United Nations, Santiago.

Goodland, R., 1997, 'Biophysical and Objective Environmental Sustainability', in A. K. Dragun and K. M. Jakobsson (eds), *Sustainability and Global Environmental Policy New Perspectives*, Edward Elgar, Cheltenham.

Hammond, A., 1998, *Which World? Scenarios for the 21st Century, Global Destinies and Regional Choices*, Island Press, Shearwater Books, Washington, D.C., Covelo, California.

Jordan, A., and H. Voisey, 1998, 'Institutions for Global Environmental Change', *Global Environmental Change*, vol. 8, no. 1, pp. 93–7.

Meppem, T., 2000, 'The Discursive Community: Evolving Institutional Structures for Planning Sustainability', *Ecological Economics*, vol. 34, no. 234, pp. 47–61.

Meppem, T., and S. Bourke, 1999, 'Different Ways of Knowing: A Communicative Turn Toward Sustainability, *Ecological Economics*, vol. 30, pp. 389–403.

Meppem, T., and R. Gill, 1998, 'Planning for Sustainability as a Learning Concept', *Ecological Economics*, vol. 26, pp. 121–37.

Modak, P., and A. K. Biswas, 2001, *Conducting Environmental Impact Assessment for Developing Countries*, Oxford University Press, Delhi.

OECD, 1979, *Interfutures, Facing the Future, Mastering the Probable and Managing the Unpredictable*, Organization for Economic Cooperation and Development, Paris.

Osborne, F., 1948, *Our Plundered Planet*, The Conservation Foundation, New York.

Raskin, P., G. Gallopin, P. Gutman, Al Hammond, and R. Swart, 1998, 'Bending the Curve: Toward Global Sustainability', A Report of the Global Scenario Group, Stockholm Environment Institute, PoleStar Series Report No. 8, Stockholm.

Serageldin, I., T. Husain, J. Martin-Brown, G. Lopez Ospina, and J. Damlamian (eds), 1998, *Organizing Knowledge for Environmentally and Socially Sustainable Development*, Proceedings for a Concurrent Meeting of the Fifth Annual World Bank Conference on Environmentally and Socially Sustainable Development, Co-sponsored by UNESCO and the World Bank, World Bank, Washington, D.C.

Söderbaum, P., 1998, 'Economics and Ecological Sustainability: An Actor-Network Approach to Evaluation', in *Evaluation Planning*, Kluwer Academic Publisher, The Netherlands.

————, 2000, *Ecological Economics: A Political Economics Approach to Environment and Development*, Earthscan Publication Ltd, London.

Tolba, M. K., 1982, 'Development without Destruction', Address to Chelsea College in 1976, in *Development without Destruction: Evolving Environmental Perceptions*, Tycooly International, Dublin.

————, (ed.), 1988, *Evolving Environmental Perceptions, From Stockholm to Nairobi*, United Nations Environment Programme, Butterworths, London.

Tortajada, C., 2000, 'Environmental Impact Assessment of Water Projects', *Water Resources Development*, vol. 16, no. 1, pp. 73–8.

United Nations General Assembly (UNGA), 1987, Resolution 42/186, Environmental Perspective to the Year 2000 and Beyond, 11 December.

WCDE, 1987, Our Common Future, Report of the World Commission on Environment and Development, Oxford University Press, Oxford.

2

Sustainable Development: A Flawed Concept

Morris Miller

'Sustainable development is development that meets the needs of the present without compromising the ability of future generations to meet their own needs.'

> *Our Common Future*, The World Commission on Environment and Development (WCED), Oxford University Press, 1987, p. 43.

'Sustainable development is ultimately a frustrating idea (when one tries) to turn it into a usable concept. But as a broad goal, sustainable development is useful. Like many important ideas, it is better than nothing for as long as there is nothing better.'

> *The Economist*, in an editorial titled, 'Inheriting the Earth', September 1992.

Nothing Better? Would a Rose by Another Name Smell Sweeter?

What we are referring to here is a phrase identifying an objective for a set of policies that focus on the relationship of development and the environment in the broadest sense of both those terms. In principle, there would not appear to be any reason to make an issue of the adoption of this terminology; it seems to accord with the traditionally accepted objective of public policy of enhancing the global public good *over the long term*. There are, however, reasons for suspecting that the sudden popularity of the change in usage from 'development as though the long-term future mattered' to 'sustainable development' is not as innocent in its origins and in its impact as it might at first appear. Indeed, from the point of view of policy-making, there are reasons to believe that the newly-minted phrase is a flawed concept and, as such, not only unhelpful—which suggests neutrality—but damaging in its impact on designing and implementing effective policies of environmental management on the national and global scales. This assessment can be gleaned from a review of the history of the use and meaning of this concept, of its success and its shortfall, and the reasons for its lack of success in furthering the 'environmental cause', whose supporters first introduced the concept with this specific phraseology only a few years ago.

The concept of sustainability is, of course, not new and has long been explicitly associated with the management policies of the ministries responsible for fisheries and forests and other natural resources. However, the use of the word 'sustainability' in conjunction with the development process itself is relatively new. The first reference to the environment in a development context dates back to 1948 when Osborne, the founder and former president of The Conservation Foundation, wrote, in a book titled *Our Plundered Planet*, that the US should pursue 'the kind of development that can be sustained'.[1] It is, he wrote, 'the kind of development that makes sense'. Sensible or not, for many years there did not appear to be any uptake of the two words in combination either by governments or non-governmental organizations or other authors. There were statements linking the two as, for example, the 1971 United Nations document called the Founex Report, which noted that 'the recognition of environmental issues is an aspect of the widening of the development concept' and in 1976, a speech by the Executive Director of the United Nations Environment Programme (UNEP), Mostafa Tolba, that contained the statement, *'a new kind of*

---

[1] For a discussion of the history of the use of the concept of sustainable development, see the chapter in this book by Cecilia Tortajada, 'Sustainable Development: A Critical Assessment of Past and Present Views'.

development is needed . . . (that avoids) such serious environmental damage as to make development simply not sustainable'. It was only in 1981 that the two words were melded into a phrase by a former President of the World Bank, A. W. Clausen, in a speech titled, *'Sustainable Development: The Global Imperative'*. The linkage of the words, 'development' and 'sustainability' was made only on rare occasions over the next few years until, in 1987, a terminological phenomenon occurred, with a suddenness that has rarely been witnessed, when the phrase 'sustainable development' virtually displaced the traditional phrase 'development as though the long-term future mattered'. The watershed was the report of the World Commission on Environment and Development (WCED) (otherwise known as the Brundtland Commission after its chairperson) published in 1987 as a book titled *Our Common Future*, which became an almost instantaneous bestseller around the world. The extraordinary event is to be found in the rapid and widespread use of the two-word phrase 'sustainable development' by governmental agencies and non-governmental organizations to relabel themselves but also to attach the phrase to the title of their programmes and events as, for example, the United Nations Conference held in Johannesburg in 2002 that was titled 'Conference on Sustainable Development'. By the early 1990s, as Cecilia Tortajada points out in her recounting of the history of the phrase, 'sustainable development became "the" paradigm of development . . . with nearly all governments embracing (it) . . . with no serious discussion ever taking place as to how the concept could be operationalized'.

Terminological changes in the form of new labels on ideas and related programmes may not appear to be of much consequence when, to all appearances, the labels mean the same thing. This would seem to apply to the substitution of a neat phrase 'sustainable development' for an awkward one, namely 'development as though the long-term future mattered'. Why then make a fuss? The question can be framed a different way: why was there a very sudden uptake of a new phrase to identify a traditional and familiar way of identifying an objective of policy-making? It is worthwhile to make an effort to answer this question, in as much as in answering it we should be in a better position to judge (a) whether the idea of attaching a new label for an old objective has been an 'innocent' process, that is, its promotion has evolved without the aid of those in positions of power abetted by a politically and philosophically supportive media, and (b) if it has been so promoted, of course with subtlety, whether that support has a hidden agenda that favours the prevailing economic system and, in doing so, those in positions of power.

From the analysis of the impact of the terminological change, we could be in a position to assess whether the incredibly popular usage of the term has been helpful or harmful for the global society-at-large or whether it has been helpful for the special interests who desire to maintain the *status quo* of the prevailing economic system—albeit slightly modified—in their own interests. The change in terminology has likely been more harmful than helpful in so far as it has narrowed the focus of the struggle for a better environment that needs to be integrated with a global development process by which the poorest half of the world's population would be enabled to rise out of dire poverty. That equalizing attribute of a development process would make the most significant contribution to the legacy that could be left for future generations.

The next section considers the issue of whether the introduction of the new terminology has helped to alter the manner of development in any part of the world to a degree significant enough to accord the maintenance of environmental quality as high a priority as that given to the speed of growth in the aggregate and to the manner of growth that would further the objectives of equity and other attributes that are the desired characteristic of a humane world. In the following section of the paper we discuss the reasons why the new logo of 'sustainable development' has become so popular and, at the same time, so ineffectual in moving the environmental/equity agenda forward, but, at the same time, effectual in giving the illusion of progress as rhetoric has been enabled to triumph over reality. In the final section, consideration is given to the issue of what should be done starting with the discouragement of the use of the 'sustainable development' logo that represents a distinctive brand of the development/ environment nexus. Hopefully the reasons for making this recommendation would then be understood and, accordingly, also the reasons for making this fuss over the issue of terminology.

The Gap between Rhetoric and Reality: The Failure of the 'Sustainable Development' Logo to Advance the Cause

Complicated policies and arguments have little place in political discourse. The public has neither the background nor the patience to digest a complicated message, so the 'simplicity constraint' makes it more difficult to put together politically appealing reforms which are Pareto improvements. For academics, this is a hard pill to swallow: we pride ourselves in the subtlety of our arguments, not in their obviousness, in the cleverness of our solutions, not necessarily in their simplicity.

Joseph Stiglitz, 'The Private Uses of Public Interests: Incentives and Institutions', *Journal of Economic Perspectives*, Spring 1998.

A hard-headed assessment of the political scene over the past decade with respect to reconciling development with the maintenance of a healthy environment would reveal that the new terminology has not been successful in moving the environmental agenda forward, let alone, as fast and as far as its sponsors and supporters had expected and hoped. For many politicians, the media, and many others, the faith in the efficacy of the relabelling of the development/environment issue resided in the appeal of the convenience of 'sustainable development' as a neat phrase. The fact that this morphing process expunged the key words 'long-term future' did not seem to matter, especially as the meaning seemed clear. There was no claim by those who coined the new phrase or by those who used it so readily that the change in labelling was due to new facts or new insights from better analysis about the emissions of pollutants into the atmosphere, except for one thing: the introduction of the 'bogey-man' factor in the form of a 'doomsday scenario', as the emissions of what were called 'greenhouse gases' were said to pose a threat to life on earth by virtue of their effect on the climate.

For the authors of the new phrase, this shift of terminology was needed to focus attention on the fact that the prevailing development policies and practices of almost all governments and international development institutions were not designed to accord a high enough priority to mitigating the harmful global environmental impacts of the development process. The quick adoption of a new terminology could logically be attributed to a widespread impatience with the slow pace and weak implementation of a full-scale environmental programme on a global scale and, to a lesser extent, on a national and regional scale. Accordingly, there was a need to introduce a shock factor. True, there had been wake-up calls in the early 1980s with the occurrence of a series of environmental disasters that were dramatic enough to have the names of their locales etched in the public's mind: Three Mile Island, Chernobyl, Bhopal, and others. With the attention span of both the public and the politicians being short and the immediate impact being geographically local and sporadic, the resulting widespread consternation did not translate into much action. In any case, dramatic environmental tragedies did not do much to stir the emotion of fear about an even more important phenomenon, namely, the slow insidious decline in the availability of land, air, and water as garbage disposal dump sites for pollutants.

An answer could be to focus on the threat of global climate change as a portending catastrophe of planetary proportions. Placing the environmental issue in the context of 'the very future of the Earth' and 'our future prospects' provides considerable drama—and pressure for action, even if that action is only identified as the symbolic gesture of signing an international protocol.[2] This would drive home the simple message about the

---

[2] A typical commentary is provided by Maurice Strong who headed the UN Conference on Environment and Development in 1972 in Stockholm and has been a 'senior adviser' to Kofi Annan, the UN's Secretary-General

imperative need to slow the build-up of greenhouse gases that were identified as the one set of pollutants that posed an end-of-life-on-the planet kind of threat. The beauty of this tactic was that the rest of the environmental agenda could piggy back on this issue. Thus 'sustainable development' became the banner under which the fearmongers marched with slogans that conveyed the simple message that the current manner and speed of development had ominous implications for everyone on the planet. Or, to put it in another way, the message was that the price of economic and financial progress on a global scale would be too high if the prevailing pattern and speed of development achieved by the present generation left a shameful legacy to future generations.

This legacy concept has proven to have great appeal, but it is important to note that this was not a new theme. In traditional environmental management policies, the key element of that legacy was generally identified in terms of the amount and quality of natural and man-made resources available for the use of future generations, including among the resources the still-remaining capacity of land, air, and water to absorb pollutants, that is, to act as garbage receptacles and thus to enable the maintenance of the quality of the environment. What was new was the shift of focus to two of the attributes of what constituted the legacy of a healthy environment, namely, the range of temperature variation and the stability of weather

---

and James Wolfensohn, the World Bank's President. Urging Canadians to support the signing of the Kyoto Protocol, he writes, 'surely in a matter as important as the very future of life on Earth, we cannot wait (to sign on) . . . Our place in the world and our ability to protect our interests and our future prospects depend on the influence we can have in co-operation with others through the multilateral framework that Kyoto provides' *The Globe and Mail* (Toronto), 'Don't blow it, Canada', 6 December 2002. (It might be noted that Canada accounts for only 2 per cent of global carbon emissions, but it is the example that is said to count to assure 'the very future of life on Earth.')

patterns. It did not seem to matter that all the proposed actions to tackle these two additional environmental 'bads' were virtually the same elements of the traditional environmental management programmes that have long been advocated and, to varying degrees, been implemented. This shift in focus did have one apparent effect in that it succeeded in putting a new label on old bottles as 'sustainable development' became the title of conferences and meetings, task force reports, and countless books, articles, and pronouncements and even of government ministries and departments and non-governmental organizations and of their programmes. There was, however, little to show for this enthusiasm for name-change beyond raising the decibels of the rhetoric of fear and the setting of specific environmentally-related targets for the near-term future without spelling out how these targets were to be achieved.

Despite the lack of progress and that of a road map to the desired future, there has been great public support for the environmental cause in vocal terms through demonstrations and in the purchase of publications devoted to the environment/development theme.[3] Surveys of the views of the Americans, Australians, Canadians, Europeans, and Japanese began to

---

[3] A sampling of the titles to be found in bookstores in the early 1990s illustrates this phenomenon:

Rescue the Earth! Conversations with Green Crusaders; The Eco-Wars: True Tales of Environmental Madness; The Fragile Environment; The Greenhouse Trap: A World Resources Guide to the Environment; On the Brink: Endangered Species in Canada; Endangered Species: The Future for Canada's Wildlife; The Global Ecology Handbook: What you can do about the Environmental Crisis, Ozone Crisis: The 15 Year Evolution of a Sudden Global Emergency; Population Explosion: Why Population is our #1 Environmental Problem, The Wasted Ocean; High-Tech Holocaust; Discordant Harmonies: A New Ecology for the 21st Century; When Technology Wounds: The Human Consequences of Progress; Making Peace With the Planet; Green Future, The Forest for the Trees? Government Policies and Misuse of Forest Resources; Blueprint for a Green Economy, Sustainable Development: Economics and Environment in the Third World.

consistently indicate that the majority of their population were more concerned about shortfalls in the implementation of 'environmental management' than they were about such issues as growth, inflation, unemployment, and poverty. There was also—not coincidentally—a significant rise in the global percentage of the non-governmental organizations, numbering more than 23,000 that took up 'the environmental cause' in its global dimensions (French 2000).[4] The public's understanding of the ramifications of the so-called 'environmental issue' was hardly deepened by the tactics of the media and of the environmental non-governmental organizations, which, as *The Economist* (December 27th, 1997) observed

have a vested interest in supporting the most alarming versions of every environmental scare (since their) own incomes, their advancement, their fame and their very existence depend (on this tactic).

The author then commented favourably on a statement by H. L. Mencken, to the effect that

the whole aim . . . (of the politicians and the leadership of non-governmental organizations) is to keep the populace alarmed—and hence clamorous to be led to safety—by menacing it with an endless series of hobgoblins, all of them imaginary.

The nub of the difficulty would appear to be that despite the effort over the 1990s until the present to instil a sense of fear about the issue of 'sustainability', the overwhelming percentage of those who declare themselves to be supporters of 'sustainable development' have only a very vague idea about the meaning of the term and, most particularly, its implications with regard to the wide and deep range of political, social, and economic changes that such support implies, giving rise to a wide gap between word and deed. In the developing countries, not much is done about the issue of the environment since the issue hardly

---

[4] According to French (2000), the number has increased from 2 per cent of this total in 1953 to 14 per cent by 1993, and it is likely to be much higher by now.

registers for the average citizen for the simple reason that, by virtue of their poverty, they have a pathetic degree of choice in their mode of living as sheer survival is a day-to-day challenge. This is especially so for those who live in the rural areas and are dependent mainly on subsistence agriculture for food and on forest scraps as fuel for cooking and heating.

It would be unrealistic to expect that people can do much about the environment unless and until they are able to escape 'the poverty trap' now, but also for their future development and, thus, for their communities. Political leaders may talk of promoting a type of development that is 'sustainable', but the question arises: what is it that is to be sustained, development or underdevelopment?

The weak follow-through on all the rhetoric has raised doubts about the sincerity and resolve of the supporters of 'sustainable development'. Such doubts are compounded by the disconnect between the characterization of the scale and urgency of what is required if an end-of-civilization denouement is to be avoided and the modest targets that have been set. It would seem the targets are set relatively low in recognition of 'political realism'. But what is, perhaps, more significant, the follow-up to all the large-scale UN conferences that were staged over the past two decades has been very weak. The targets and 'plans of action' have been mocked. This weak follow-up is illustrated by one glaring example: the Johannesburg conference's 'plan of action' contained proposals regarding such key measures as the shift to greater reliance on environmental-benign energy sources. These proposals were almost identical to those articulated 20 years earlier in the 'plan of action' of the 1981 UN Conference on New and Renewable Sources of Energy that was held in Nairobi—at which time, it should be noted, the issue of global climate change with its warming trend was not even on the agenda, nor mentioned as a serious problem in the voluminous studies that were prepared for the conference.[5] It is worth considering the reasons for the

---

[5] As the Assistant Secretary-General of that conference in Nairobi, this example of the lack of progress stands out as an illustration of the futility

fact that there is so little to show for the efforts of the so-called 'environmental movement' or 'greens' that has been marching under the banner of 'sustainable development'.

The Ambiguity and Operational Vacuity of 'Sustainable Development' as a Concept

Every environmentally aware politician is in favour of 'sustainable development' but what on earth does the phrase mean?

The Economist, September 1989

The ambiguity of the phrase that identifies 'sustainable development' as an objective of policy begins with its definition. On the face of it, 'sustainable development' seems a simple enough concept. Talking about the meaning of the concept, Robert Solow remarked,[6]

it is not clear (to me) that one can be more precise than that we, the present generation, should not satisfy ourselves by impoverishing our successors (that is) sustainability is an obligation to conduct ourselves so that we leave to the future the option or the capacity to be as well off as we are.

That commonsense definition is so general that it hardly sheds much light on the reasons for the term having had such a widespread appeal and what it implies in terms of policies and programmes, let alone institutional changes of a systemic nature. Those who have attempted to be more precise in their interpretation of the definition have found that they were entering into a hornet's nest of controversy as to what the

---

of large-scale conferences in terms of actionable proposals that are, indeed, activated. For a discussion on this issue, see Miller (1994).

[6] The reference to Solow's views is from his lecture titled 'Sustainability: An Economist's Perspective' delivered in 1991 at Woods Hole Oceanographic Institution. He added that he thought it was difficult to apply the principle to the problem of saving for the benefit of future generation. This is taken up later in this paper.

definition means and whether it has any operational merit. For some the definition assumed the status of the holy grail no matter what it meant, and for others who were skeptics or critics—including myself—it has been characterized as an oxymoron, a word defined by the American Heritage Dictionary of the English Language as a 'rhetorical figure of speech or writing in which the epigrammatic effect is created by the conjunction of incongruous or contradictory aims'[7] or, more charitably, 'vacuous', which the dictionary defines as 'devoid of meaning; synonym: empty'. The unflattering characterization of 'sustainable development' as an oxymoron that implies emptiness or being devoid of meaning is perhaps too weak a condemnation when confusion and false assumptions about each of the meanings of these words and their relationships confound effective policy-making.

The confusion is evident on an examination of the range of definitions put forward by reputable economists and others. As early as 1992, John Pezzey, in a World Bank publication titled 'Economic Analysis of Sustainable Growth and Sustainable Development' (Environment Working Paper No. 15), listed definitions of 'sustainable development' compiled from many sources. This prompted one commentator who reviewed Pezzey's manuscript to make the obvious comment that 'sustainable development is one of those catchphrases that mean different things to different people'. And the list of definitions has since grown a lot longer. The result of this plethora of definitions and their seeming contradiction in meaning, scope, and implications for action is that the public understanding about what is involved in its pursuit is very weak, both as to what the end goal looks like and about how to be involved as a citizen taxpayer, organizational supporter, and/or as a voter. The ambiguity and complexity can be illustrated by a few examples beyond the simple commonsense comment of Professor Solow cited above.

---

[7] The Oxford English Dictionary goes further in its definition of oxymoron, adding, 'literally self-contradictory or absurd'.

In his article titled 'Towards a Sustainable World', David Pearce (1996), stated that sustainability would call for 'investing in the environment to ensure that the stock of environmental assets is not reduced overall'. In another journal article titled 'Sustainable Development: An Overview', he interpreted this phrase to mean

leaving the same or an improved resource endowment as a bequest to the future . . . (that is) the total stock of all forms of wealth (including environmental wealth) must not be depleted.

Robert Repetto (1989), differed by stating that

sustainable development does not mean the preservation of the current stock of natural resources or of any particular mix of human, physical and natural assets (since) as development proceeds the composition of the underlying asset base changes.

Robert Solow (1992) entered into this terminological fray on the side of Repetto when, in a lecture presented at the fortieth anniversary of a Washington-based non-governmental organization named Resources for the Future, he observed that

the duty imposed by sustainability is . . . not to consume humanity's capital in the broadest sense . . . (and in that way) bequeath to posterity, whatever it takes to achieve a standard of living at least as good as our own.[8]

His definition provides an excellent illustration of the slippery nature of the concepts of 'sustainability' and of 'development' and of their relationship to each other when they are used to provide a basis for policy-making. This slipperiness can most clearly be shown by trying to determine what is meant by the objective and by the actions that are called for.

With regard to the objective of achieving 'a standard of living at least as good as our own', we need to seek clarification about

---

[8] Solow (1992) on the subject of sustainability long predate the introduction of the phrase, 'sustainable development'. The first of his articles to focus on the subject with the word 'sustainability' was 'Sustainability: An Economist's Perspective' for the Woods Hole Oceanographic Institution.

the point of reference suggested by Solow. It would be a dubious achievement if one takes that to mean a standard of living enjoyed by those living in our prevailing economic, financial, and political system where half the inhabitants on this planet earn less than $2 per day, with all that that extreme deprivation implies. If the reference to achieving 'a standard of living at least as good as our own' is meant to suggest that *all* of humanity achieve the average living standard of those in the richer industrialized part of the world, we come up against impossibly severe environmental limits when the richer 20 per cent of humanity politically and socially for those richer 20 per cent of humanity who now utilize about 80 per cent of the capacity of the planet's air, water, and land. Those who use a very disproportionate percentage of the absorptive capacity of the planet's air, water, and land as a garbage receptacle cannot be expected to accept this objective when it implies so radical a change in lifestyle.

With regard to the stipulated actions that are called for, namely, 'consume humanity's capital' and '(do)whatever it takes', we need to also seek for clarification about the operational guidelines. For example, would it be along the lines of 'the Rule' elaborated by Hartwick (1977) that pertains to how much needs to be saved from the use of natural resources that would be depleted and that then needs to be invested in reproducible capital so that the total returns could be sustained over time?[9] The difficulty as to what this implies operationally is not hard to imagine when the very concept of a 'natural resource' is very ambiguous, that is, its definition itself very much depends on technological change and the ways in which the various economic systems use and could use these technologies, which, in turn, is dependent on other variables. Or to cite another example, do we follow 'the rule for achieving sustainability' set forth by James Roumasset, that can be characterized as gobbeldygook:[10]

---

[9] Hartwick, (1977). This issue is elaborated in Anand and Sen (1996).
[10] Roumasset (1990, p. 40).

in order to maximize the sustainable consumption of the future, broadly defined to include environmental amenities, the rule is to deplete resources and accumulate capital until the rate of return of saving a unit of a resource is equal to the rate of return on capital, (a rule that) calls for more discrimination in distinguishing between efficient and wasteful uses of the earth's bounty. But depletion is not a sufficient metric of waste or even non-sustainable use (as) . . . waste is not the necessary consequence of economic development but of inappropriate institutions, infrastructure and economic policies (that adhere to) the central criterion for using resources in a sustainable manner (which is) to deplete resources stocks only when the contribution these resources make to current income, including capital formation, is greater than the opportunity cost in terms of future benefits foregone (that) does not mean depletion per se should be interpreted as non-sustainable.

All this scarcely comprehensible jargon and the troubling questions leave unanswered how we would determine such things as 'the rate of return on capital', 'efficient and wasteful use', 'depletion', and especially 'the opportunity cost in terms of future benefits foregone' and 'the rate of return of saving a unit of a resource'. In effect, the theoretical literature on the 'optimal rate and manner of growth' provides little help to those who have the responsibility to frame appropriate developmental and environmental policies. To top it off, the policy-maker is left virtually clueless as to how to proceed in the real world when the frontiers of sustainability are shifting so fast and so far thanks to the unprecedented pace of scientific and technological advances in what is referred to as 'the information age' or 'the knowledge economy'. The fact is that human ingenuity in science and technology and in social and cultural adaptation is continually moving the frontiers or limits of economic possibility thereby, precluding a precise definition of 'sustainable development' in real contextual terms that are of relevance for policy-makers. In brief, we are left virtually clueless as to how the term 'sustainable development' could be—or should be—interpreted for use in the policy-making process. Sceptics and critics of the 'sustainable development' concept simply claim that it falls short or fails on grounds of logic and applicability in the real world.

It might be pertinent to note what these sceptics and critics have to say. Terence Corcoran, wrote this about the concept:[11]

never have two words been used so much with so much inconsistency . . . For the most part, nobody seems to care what the words mean, or whether they even have any real meaning . . . It is fast becoming a landfill site for every environmental idea. . . . Have we reached a point where 'sustainable development' has become a hazardous concept?

Other critics of the concept of 'the sustainable development movement' make similar charges—though in a more temperate manner—to argue that not only is the term devoid of meaning and operationally vacuous, but that the use of the term as an oft-repeated phrase is very injurious in as much as it has thereby become a cliché, overused in public discourse to the point where it has become irksome and without any meaningful meaning.[12] A sampling of the commentary of such critics would indicate why the phrase is believed to be unhelpful, immoral, and an oxymoron, among other epithets.

Herman Daly has expressed strong doubts about growth being compatible with the maintenance of environmental quality. In his view, the concept of 'development', if it implies change would be acceptable, but would be unacceptable if development was to be defined as expansion of the global economy: growth should then be regarded as an 'economic oxymoron'. He assails the Brundtland Commission for accepting

[11] 'Sustainable Development: A Dumpsite for Ideas', *The Globe and Mail* (Toronto), 23 March 1990.

[12] *The New York Times Manual of Style and Usage* defined a cliche as 'words or phrases that are all things to all men, the good ones lending a helping hand but adding insult to injury, the bad ones being beneath contempt'. This is taken from an interesting column by Robert Fulford in *The National Post* (9 July 2002) in a column titled, 'Giving clichés short shrift: they are a crime against language'. He cites *The War Against Clichés* by Martin Amis, who has charged that 'they deaden prose and also information, discussion and the people who use them, and they limit and enclose thought, forcing it down predetermined channels'.

the premise of global growth by a factor of five to ten that, he believes, 'would move us from unsustainability to imminent collapse', and notes that

sustainable growth is the buzz word of our time (but) is hollow political verbiage, totally disconnected from logical and physical first principles when the aggregate economy is assumed to grow forever (Daly and Cobb 1990).[13]

This puts the emphasis on the environmental limits to growth, that is a variant of the Malthusian thesis (about which more later).

This dismal view of 'development as growth' has also been expressed by Ezra J. Mishan, who has been the most articulate spokesman for a school of thought that challenges the facile assumption that development in the form of growth is even desirable, let alone sustainable. Given not only its adverse impact on the environment but, more particularly, its adverse impact on a host of social and cultural attributes that constitute a 'civil society', he is sceptical about any concept of growth that could court 'ecological disaster' and that is treated as synonymous with progress in human well-being, at least in the industrialized countries that he refers to as 'the West'.[14]

Speaking not so much as an economist but as an ordinary citizen, there are several considerations that prompt me to doubt whether further economic growth in the West has much, if anything, to add to human happiness or to the good life . . . Several other considerations suggest that further economic growth is likely to be positively detrimental to human welfare .

[13] Daly and Cobb (1990) define the desirable and necessary objective as the attainment of a condition that is called 'a steady-state', that is, a state of affairs where growth stops when it reaches 'an optimal scale relative to the ecosystem, that is, the economy should be considered a steady-state subsystem of the environment . . . (Yet) the concept of optimal scale of the aggregate economy relative to the ecosystem is totally absent from current macro theory'.

[14] Mishan has written several books on welfare economics (Mishan 1969a, b, c). Among his professional articles, the most relevant one for our purposes is Mishan (1993) from which the quote has been taken.

. . (for to) support a policy of drifting along with the current economic momentum is tantamount to a posture that implies a willingness to run some risk of ecological disaster at a time when, to say the least, there is no presumption of any significant gain in welfare from further economic growth.

Another critic, Michael Redclift, writing in his book, *Sustainable Development: Exploring the Contradiction* (1987),[15] concurs that the concept of development needs to be defined in a particular way if, in talking of sustainable development, we are to avoid 'the substitution of moral convictions for thought':

the constant reference to 'sustainability' as a desirable objective has served to obscure the contradictions that 'development' implies for the environment . . . . Environmental change should be understood as being a social process, inextricably linked with the expansion and contraction of the world economic system . . . . Therefore, 'development' must be subjected to redefinition since it is impossible for accumulation to take place within the global economic system we have inherited without unacceptable environmental costs. Sustainable development, if it is to be an alternative to unsustainable development, should imply a break with the linear model of growth . . . that ultimately serves to undermine the planet's life support system.

Ignacy Sachs of the L'École des Hautes Études en Sciences Sociales makes much the same point about the inherent contradictions in the narrow interpretation of the concept of 'sustainable development' that is current in political circles:[16]

Once again the politicians have seized on the language of sustainable development while emptying it of meaning, replacing any reference to economic growth with the term 'sustainable development' as the situation in social terms has been greatly deteriorating over the last five years . . . It can never be said too often: there can be no sustainable development so long as the social crisis persists . . . . The path to the future seemed laid out with the heads of state committing to Agenda 21 at the Rio conference where

[15] p. 199. His book explores the theme of how the development process poses problems for sustainability under capitalism with its many inequitable features.

[16] Sachs (1997).

the new paradigm of sustainable development was accepted. Five years later, it is time to drop our illusions . . . .

Wilfrid Beckerman, writing in an article in *World Development* (1992) titled 'Economic Growth and the Environment: Whose Growth? Whose Environment?', states that, in addition to being devoid of operational value, 'the concept of global "sustainability" that is so widely encountered these days is immoral'. His quarrel is with the juxtaposition of concern for *inter*-generational equity over that of *intra*-generational equity, a position akin to that of Ignacy Sachs and many others who deprecate the practice of treating environmental policy issues as a matter deserving priority ahead of policies focused on achieving social objectives, including poverty alleviation and 'good governance', without which no policies for human betterment could succeed over the long term.

Recognizing this line of criticism, the Administrator of the United Nations Development Programme (UNDP), James Gustave Speth, relabelled 'sustainable development' as 'sustainable *human* development'. In 1994, UNDP published a paper titled '*Sustainable Human Development*—From Concept to Operation: A Guide to the Practitioner' (Banuri et al. 1994), in which a valiant effort was made to move beyond the original definition enunciated in 1987 in the Brundtland report. According to this paper,

Sustainable human development (SHD) . . . is a new development paradigm that is more than simply the sum of human development plus sustainable development as it brings to the development agenda the need for special attention to social capital, i.e., voluntary forms of social regulation. SHD can, therefore, be defined as 'the enlargement of people's choices and capabilities through the formation of social capital so as to meet as equitably as possible the needs of current generations without compromising the needs of future ones'.

This broadening of the original concept from two to three words does, of course, add a dimension that is highly commendable in calling for redistribution from the rich to the poor and for participation by the poor. Tying the word 'human' to the concept of sustainability does, of course, introduce the idea of enhancement of the capabilities of the population to lead more fulfilling lives and, thereby, raise their ability to generate higher incomes both now *and in the future*. But this effort to add the social element as a tag-on to environmental policy only succeeds in compounding the confusion as to the meaning of 'sustainable development'. Much more importantly, it also raises the question as to whether or not it would be preferable that environmental management policy be treated in the context of development policy as only one component among many, though admittedly a very important one, and that development policy treat the issue of global growth in a manner that gives due weight to both the short-term and the long-term future of a world that would merit the appellation 'civilization'.

The upshot of all this is that it is extremely difficulty—if not impossible—to comprehend the concept of 'sustainable development' in a *meaningful* fashion when it means different things to different people. The concept does, however, convey a positive idea. 'What could be wrong with it?' is the usual response even from those who recognize its ambiguity and complexity and non-operational attributes. Thus 'sustainable development' may win support, but, as noted before in connection with the lack of progress of those who march under the banner of sustainable development in moving their cause forward, one has to ask: support for what in particular? When everyone is for a programme without knowing what it means, the enthusiasm is not likely to be translated into support for specific programmes and projects and for the necessary changes in institutions and policies and programmes that would be required. That suits the vested interests who wish to minimize any institutional changes to a tee: 'fine to talk the talk if it obviates the need to walk the walk'.

'Sustainable Development' as a Deflector from the Institutional Limitations of the Prevailing Global System

Environmentalism (as a movement) has been ameliorative and corrective— not a restructuring force . . . Responding successfully to the multi-faceted global environmental crisis will (therefore) be a difficult political enterprise . . . . Can we move nations and people in the direction of sustainability? Such a move would be a modification of society comparable in scale to only two other changes: the agricultural revolution of the late Neolithic and the

Industrial Revolution of the past two centuries, both of which were gradual, spontaneous and largely unconscious . . . . If we actually do it, the undertaking will be absolutely unique in humanity's stay on the earth.' William Ruckelhaus, 'Towards a Sustainable World', *Scientific American* Special Issue: Managing Planet Earth, September 1989.

Over a decade ago, William Ruckelhaus, a former director of the U.S. Environmental Protection Agency, made the above statement during the period of the first surge of the use of the term 'sustainable development' but talked not of 'greenhouse gas emissions and climate change' but of 'the multi-faceted global environmental crisis'. Well before the new terminology took hold, the narrow focus on greenhouse gases and the dynamics of temperature levels as *the* global threat had begun to ring alarm bells. At the time Ruckelhaus was not alone in talking of change in radical terms; other 'eminent personages of the environment movement' declared that what was needed was the following:

- 'The needs and aspirations of today could be reconciled with those of tomorrow providing there are fundamental changes in the way nations manage the world's economy in human behavior, economics, politics or institutions of government' (MacNeill 1989).

- 'The challenge we now face is nothing less than that of creating a whole new approach to the goals of growth, to the processes of growth and to the systems of incentives and penalties which determine our patterns of growth (and this, in turn implies that) the real alternative to "no growth" is "new growth", (that is), a new approach to growth both in the industrialized and developing societies, the revamping of the present system of arrangements and institutions to better serve the interests and aspirations of the developing world." (Strong 1977, p. 10).

And then there are the voices of the media joining the chorus and abetting it. This process of fanning the flames of concern and identifying the challenge in apocalyptic terms is typified by the comment in 1989 of John B. Oakes, who talked of environmental deterioration posing 'an even greater threat to life on this planet than the nuclear threat that it encompasses'. He goes on to say that authoritative sources are alleging that 'environmental changes are putting the future welfare of human society at risk.... This is a war too important to be left to the politicians.' He did not include greenhouse gas emissions on the list of those very troubling environmental concerns![17]

The dramatic choice of language in these messages could not have gone unnoticed, especially as these views were those of persons who could hardly be categorized as politically radical. It indicated that there was, indeed, a serious undercurrent of protest which, if neglected, could pose a threat to the established order of things, But there need have been little or no anxiety about the reverberations of these clarion calls for radical systemic institutional change since the calls were not accompanied by a commensurate political programme of action, and, more importantly, since every democratic political/economic system has built-in self-defense mechanisms that make systemic change very difficult to achieve.[18] This defence takes two forms: (i) the constraining power of political correctness, and (ii) language change to soften the rough edges of the prevailing system. These are worth examining to explain why an ambiguous and vacuous terminology became desirable from the perspective of those who sought to minimize the pressure to do what might have been necessary if the challenge was to be addressed in a forthright and effective manner.

The Constraining Power of Political Correctness in Advancing Proposals for Action

William Ruckelhaus (1989) provides a good example to illustrate this point: after posing the challenge in transformational terms as 'a modification of society comparable in scale to only two other changes: the agricultural revolution of the late Neolithic and the Industrial Revolution of the past two centuries', he proceeds to

---

[17] In an editorial in the 12 January 1989 issue of the *New York Times* under the title 'Bush's Shell Game', Oakes wrote: 'This war to defend environment is too important to be left to the politicians.'

[18] See Miller (1993).

identify the key obstacle to making this transformation as 'the familiar one of externalities: the environmental cost of producing a good or service that is not accounted for in the price paid for it'. Ruckelhaus recommends that people be made to pay the 'full cost' of using a resource, thereby, in his words, 'bending the market system towards long-term sustainability', a proposal cited a dozen years later by *The Economist* (in its 6 July 2002 issue) in an editorial stating that 'the greening of the market to get prices right . . . (is one of) the three powerful forces for achieving sustainability'.[19] It remains unclear as to how 'full costs' are to be identified and measured and allocated, nor how, as part of this process, anyone could manage to implement the recommended rule of 'internalize the externalities', nor how the market is to be 'greened'.

Many others have put forward proposals that place reliance on the market process with its related incentives to induce citizens and corporations to do 'the right thing' environmentally. As a typical example, Tietenberg (1990), wrote of 'the power of the market (that) can be harnessed to economic incentive policies for the achievement of environmental goals, in effect, turning the market into a powerful ally'. One proposal is to establish a trading system in greenhouse gas pollution permits, an idea that is tantamount to giving out licences to pollute to corporations and, at the same time, expose governments to pressure by these corporations to raise the permissible level of the noxious emissions when the total permissible output of pollutants is being determined. It seems to have mattered little to those who propose this approach that environmental management is recognized to be an area of policy where 'market failure' prevails to a degree necessitating a heavy involvement of the government to ensure

---

[19] The other two are 'the empowerment of local people to manage local resources and adapt to environmental change, and the encouragement of science and technology, especially innovations that reduce the ecological footprint of consumption'. The feature story of the issue is titled, 'The Great Race'.

the promotion of what is known as 'the public good', and that this applies especially to greenhouse gas emissions, which has been characterized as 'the quintessential public good'.[20]

It can be said that in championing the environmental cause this approach remained well within the bounds of the phenomenon of 'political correctness' as none of the specific proposals poses any threat of serious change to the prevailing economic/financial system, which thrives on forced obsolescence and waste by stimulating 'wants' that go well beyond 'needs' and requires to cultivate what Ruckelhaus himself has characterized as 'the culture of unsustainability' to maintain its momentum and stability. Thus, even the very same persons who voice the need for radical systemic change have been putting forward mousy proposals for the problems that they have characterized as elephantine.

There is a long list of other proposals that have been put forward that have the virtue for the powers-that-be that they pose no serious threat to the established order of things.[21] Many of these proposals have been repeated over and over again as part of the many so-called 'plan of action' at the UN conferences in Stockholm, Nairobi, Rio de Janeiro, Kyoto, and Johannesburg. It might even be said that there has been a regression in the sense that the movement towards establishing or strengthening institutions and policies favouring a manner of development having high regard for the long term was being weakened by political leaders even as their rhetoric increased.

What is significant about almost all of the proposals put forward is that not one of them would seem to qualify as part of the 'restructuring force' and 'fundamental change' that William Ruckelhaus, Maurice Strong, and Jim MacNeill deemed necessary to meet the formidable environmental challenge. The

---

[20] See Heal (1999, p. 222).

[21] For a listing of other proposals and some amplification of what they would do in terms of raising funds from various sources and how they would be allocated to various programmes, see Miller (1994) or the website www.management.uottawa.ca/miller.

upshot is that the political establishment of the prevailing economic/financial system could feel comfortable with proposals to achieve a 'feel-good' goal like 'sustainable development' that meant different things to different people. This brings us to the second self-defence item in their arsenal.

The Use of 'Sustainable Development', to Obfuscate and to Deflect Change

In the recent past, there has been a terminological shift that has had the effect of softening the rough edges of the prevailing global economic system: that is, language has been used as an instrument of 'the establishment' to make the prevailing economic/financial order of things more politically, economically, and socially acceptable.[22]

It may be noted that the emergence of the 'sustainable development' logo and the subsequent shift of emphasis in environmental policies and programmes coincided with the tenure and the legacy of Ronald Reagan and Margaret Thatcher, who were leaders hardly noted for their support of the poor and for environmental management as they waged war on the regulatory regime that constituted a key part of environmental policy and programmes. In subtle but important ways, it is clear that their political fortunes, as well as those of other political leaders in those countries and elsewhere who had also had a low regard for the long-term future beyond their tenure in power, stood to benefit from a language change that dropped reference to 'long-term' for the ambiguous word 'sustainable'. And in dropping the phrase 'long-term', these leaders achieved two desired effects: they drew attention away from the long-term future as an element of the

---

[22] Examples abound. 'The capitalist system' has given way to 'the market system', 'firing workers' has become 'corporate down-sizing', 'recessions' have become 'corrections'; 'workers, professors and other people' have become 'human capital' or 'human resources' that crudely signifies that people are regarded as instrumentalities rather than those the economy is supposed to serve.

objective of policy by making the nature and timing of whatever it is that was promised unclear, and they drew attention away from the powerful political and economic/financial agents—'the establishment'—who, by their very nature, make decisions only on the basis of short-term criteria.

Yet, with respect to the issue of a healthy environmental future, it is the long-term aspect that poses the formidable challenge: if the global economy is to grow and grow more equitably—so that the poorer half of humanity enjoy a more rapid rate of growth of income and all that goes with it—there must be time for major adjustments to be made. Let us focus on one aspect for illustrative purposes. If the manner of growth must change for the sake of a better environment and equity, there must be a profound change in the sources and the uses of energy. This is a formidable challenge that calls for major shifts in the global energy mix which have taken a very long time to be achieved. For example, the past shift from the peak use of a fossil fuel like coal to the peak use of another fossil fuel, oil, has taken over three generations and was concurrent with a major transformation of the global economy. There is no escape from the judgment that tackling the global environmental challenge in earnest in all its manifestations entails a *mega* shift in the way factors such, as energy, are used, and in the related institutional arrangement at several levels of governance.

To achieve this transformation more rapidly and within the pain tolerance permitted by political considerations is quintessentially an issue of political will. Modest changes would help but would clearly not suffice. Recognizing this, the 'sustainable development' logo is very attractive to 'the establishment' as a diversionary and obfuscating tactic: political leaders can thereby buy into the tragic denouement hypothesis related to climate change and appear responsive to the anxiety of citizens by setting ambitious targets and signing protocols using terms that, by their very nature, are ambiguous in their meaning and operationally vacuous.

The Malthusian Hypothesis Underlying the Sustainable Development Concept

Where our ancestors feared the apocalyse as a matter of superstition or faith— echoes from the ages—we fear it as a consequence of knowledge (that) we now lie in a new age of the apocalyse that is daily announced in the media arising from authoritative sources in science, medicine and politics credibly warning of collective suicide through greed, technology or stupidity . . . (that threatens) extinction caused by the human contribution to global warming, the rape of the environment by starving masses or a 'nuclear winter' through senseless war.

William Thorsell, 'Risking it all', *The Globe and Mail*, 10 June 2002

There is another self-defence mechanism that the elites of all well-established systems have long used wherever there has been a need to explain why things are not going as they should: find a scapegoat for whatever ails the society. The use of the climate-change threat is, thus, a familiar ploy: in its essential features it takes the form of what has come to be called 'Malthusianism' that, in its essence, places the blame for societal ills and ominous future prospects on the limits of Nature rather than on the prevailing political/economic institutional arrangements as a system.

Writing at the end of the eighteenth century, Thomas Malthus put forward a simplistic doomsday model that related the differing rates of growth of population and of food supplies to demonstrate that they are periodically brought into balance by pestilence, famine, and wars. In the 'dirty' early years of the Industrial Revolution, William Pitt, England's prime minister of the time, recognized a political benefit in promoting the Malthusian hypothesis that explained why the working poor were living in misery and why, as they propagated, their ranks would be decimated periodically by famine, war, and pestilence to bring their food needs into equilibrium with food supplies. His model related the growth of population to the rise in agricultural production and proved a handy rationalization for the misery of the poor: the blame was placed on their propensity to fornicate and breed children and not the ruthless imperatives of the nascent capitalist system.[23]

Malthus wrote successive editions in which he added those facts that bolstered his simplistic hypothesis in a travesty of the scientific method: first there was the hypothesis, then selection of only those facts that supported the hypothesis! The endorsement of the hypothesis was exceptionally great among the elites at that time and periodically over the next two centuries as neo-Malthusians emerged in great numbers, the most recent dramatic instance being the publication of the the best-seller *The Limits to Growth* (Meadows et al. 1972) and the formation of a group known as The Club of Rome that was comprised of famous persons who were the sponsors of the Meadows study. The timing of publication followed 'the international oil crisis shock' of the early 1970s and its sales, therefore, can be seen as a reflection of either widespread fear for the future or an embrace by religious zealots of the idea of the coming of the Apocalypse.[24]

For the rest of us who wish not merely to avoid catastrophe but to leave a better world behind us, there are often dark clouds on the road ahead that give rise to anxiety and pessimism—but not the extinguishing of hope. Approaching the Malthusian hypothesis in a rational manner rather than in a religious or ideological manner would call for an examination of the facts with regard to trends and the interpretation of their implications. It is, therefore, instructive in this context, where we are assessing a phrase that has Malthusian roots, to revisit the 'limits to

[23] See Foster (1998) for an interesting history of the Malthusian essays and their impact over time.

[24] The feature story in the 1 July 2002 issue of *Time* is titled 'Apocalypse Now.' It includes commentary and statistics on the number of believers and the impressive sales of the books sold that dwell on the theme of Armageddon and the Apocalypse. The findings of a recent poll of Canadians published by the Canadian Press on the issue of religious beliefs revealed that over 57 per cent believed in angels, 31 per cent in extra terrestial aliens, 30 per cent in ghosts, and a majority in life after death.

growth' hypothesis to see if there are some parallels to the doomsday modelling that underlies the thesis of *The Limits to Growth*. That study concluded that the estimated known reserves of coal would run out in 110 years, natural gas in 22 years, and petroleum in 20 years, and even if there was to be a five-fold increase in the known reserves the limits of availability (and presumably, marketability) would stretch another 40 years for coal, another 27 years for natural gas, another 27 years for petroleum, and so on. The study went on to state

(if one assumed that) technological optimists are correct and that nuclear energy will solve the resource problems of the world, growth would be stopped by another constraint, namely, the limited capacity of the air and water to absorb the rising quantity of pollutants . . . . The basic behaviour mode of the world system is exponential growth of population and capital, followed by collapse . . . . Growth will be stopped by pressures that are not of human choosing.

To the extent that the Malthusian hypothesis or way of thinking is a key underpinning of the 'sustainable development' thesis, there arises the threat of a credibility gap as the estimates above have been proven wrong so often for so long.[25] Yet, we still find that this type of doomsday scenario prognostication remains a common message of governments and of non-governmental organizations that make pronouncements about environmental issues. The World Wildlife Fund (WWF) (2002) provides a recent example in its crudest form à la Meadows when it

[25] The type of model or mode of thought that produces this apocalyptic outcome is exemplified by the metaphor of the lily pond to illustrate the nature of exponential growth. The metaphor goes somewhat along the following lines: 'A pond will be filled with lilies in thirty days; the lilies double in number every day; if the decision is made to act only when the pond is half full, on what day will action be taken?'

The answer is: the 29th day. Lester Brown has written a book with the 29th day as its title (Brown 1978). The question is not whether this is good arithmetic but whether the lily pond dynamics can stand up as an appropriate metaphor for the world we live in. For more on this theme, see Miller (1998).

calculated in a recent report that there is 'a human deficit with the Earth' and, on present trends, point to a sombre outcome with astounding precision:

by 2050 humans will have consumed between 180 and 220 percent of the Earth's biological capacity . . . (and) the standard of living and human development as measured by average life expectancy, educational levels and world economic product will start to plummet by 2030.[26]

While the Malthusians have been far off the mark with regard to most of the constraints on growth, they have a strong point with regard to the limited capacity of the air, water, and land to absorb the pollutants and waste of the global economy, especially when the future rapid growth of the developing countries with the resultant pollutants is factored into the equation. One would imagine then that the 'sustainable development' movement is on strong grounds in focusing on limits and the unsustainability of continuing along the present track. But the movement has virtually abandoned this focus to operate under the rubric of 'sustainable development' with a narrow focus on only one element of the environmental degradation spectrum of 'bads' that are associated with the prevailing economic system, namely, the rising greenhouse gas emission levels and the related doomsday scenario of rising temperatures and volatile weather. In doing so, a veritable storm of controversy has erupted putting the Malthusian tag on the movement and, thereby, seriously tarnished both its motives and credibility. This adverse impact can be illustrated by the views of commentators who have drawn a conclusion about the debate on the environmental issue of global warming similar to that articulated in an article by Wente (2002):

---

[26] As reported in a WWF Press Release from their document, *The Living Planet Report 2002*. The report cites the calculation 'humans are currently running a huge deficit with the Earth, using over 20 percent more natural resources each year that can be regenerated—and this figure is growing each year'.

the real story is how science has been corrupted by the official doctrine (in a manner) that at its core is anti-intellectual . . . . I suspect that our belief in global warming is at root theological (as) the tendency to blame ourselves for natural calamities dates back to the dawn of time, . . . (a cultural phenomenon that has continued until today with) the widespread unease that our chief sin is materialism and progress.[27]

The undisputed fact is that in this matter there happens to be a great deal of controversy and the controversy revolves around the following issues: whether one can speak meaningfully of 'average temperature levels' in global terms; whether anyone is able to assert with authority that these levels are rising at a rate significant enough to be a cause of great concern; and whether human activity has—or is ever likely to have—an impact on climate that could justify the proposed measures to reduce those emissions. Without entering into the debate 'with both feet', as the saying goes, it would suffice to note that for policy purposes— and to keep our focus on the role of the factor of fear about climate change that underlies the 'sustainable development' movement— it is the last aspect regarding the role of humans as causal and remedial agents that merits emphasis. In this regard, it would be appropriate to demonstrate scepticism by quoting Philip Stott, who has been in the forefront of the debate about both the scientific underpinning and what humans can do about the phenomenon of climate change. He writes of the arrogance of

the idea that we (humans) can control a chaotic climate governed by a billion factors through fiddling about with a couple of politically selected gases . . . Climate is one of the most complex systems known, yet the idea that we can manage it by trying to control a small set of factors, namely,

---

[27] Wente, M., 'The Kyoto-speak brainwashers', *The Globe and Mail* (Toronto), 7 December 2002. She quotes from a book, *Taken by Storm* (Essex and McKitrick, 2002), that she characterizes as both irreverent and devastating in their explanation of 'the limitations of climate-change science to a scientifically-challenged public'. Essex is described as 'a senior player in the world of climate science who specializes in the underlying mathematics, physics, and computation of complex dynamic processes such as climate'.

greenhouse gas emissions, is a basic fallacy. Scientifically, this is not merely a matter of uncertainty; it is a lie.[28]

The nature and degree of the uncertainty 'as a policy problem' is succinctly set forth in an informative article titled 'The Role of Economics in Climate Change Policy' written by W. McKibbin and P. Wilcox (2002, p. 109):

Although greenhouse gases can trap energy and make the temperature warmer, and the concentration of those gases has been increasing, it is far from clear what those facts mean for global temperatures.... It is even quite hard to prove that global warming has begun.

A long list of scientific uncertainties makes it . . . impossible to say how much warming has occurred to date or how much will occur in the next century.... The cost of reducing greenhouse gas emissions is also uncertain
. . . .

From climatology to economics, the uncertainties in climate change are pervasive, large in magnitude and very difficult to resolve. *In short, uncertainty is the single most important attribute of climate change as a policy problem.*[29] (emphasis added)

Given this uncertainty, it would seem advisable to resist using the dreaded denouement as the basic premise of the 'sustainable development' movement and to resist setting targets *à la* Rio, Kyoto and Johannesburg. It is not merely a question of the

---

[28] www.globalwarming.org/polup/pol14-4-01.htm, the Global Warming Information Page. Stott in a recent article (www.junkscience.com/mar02/wsj-stott.htm) has noted that 'despite a short-term rise in temperature of around 0.6 degrees centigrade over the last 150 years, the long-term temperature remains, overall, one of cooling. It may not be too long, therefore, before we see the ice spreading again . . . . We are currently emerging—granted in a somewhat jerky fashion—out of the Little Ice Age that ended around 1880 . . . . Our current interglacial period is already 10,000 years old and no interglacial period during the last half-million years has persisted for more than 12,000 years, most having had life spans of only 10,000 years or less. Statistically, therefore, we are due to slither into the next glacial period . . . . At worst, withdrawing gases might help speed the descent into the next glacial period'. See also Stott (2000).

[29] McKibbin and Wilcox (2002).

climatic modelling being proven wrong; rather the damage of target-setting in the face of so much uncertainty would likely be great in augmenting scepticism and cynicism about the fuller range of environmental issues that need to be addressed. In a word, the process would likely backfire.

The societal damage of scepticism and cynicism as a widespread social phenomenon is enormous as both these, along with ignorance and indifference, are the key components of political inertia.[30] This being so, there is likely to be even further damage inflicted on the broad range of issues in 'the environmental cause' by supporters of the doomsday climate thesis who ask politicians and the public to rely on the argument that there is a consensus among the *majority* of scientists. They base this on the conclusion by consensus of several hundred scientists—only a small percentage of whom have expertise in climate-related science—who have been involved in preparing the reports of the Intergovernmental Panel on Climate Change that has been the basis for the proposed targets of the Kyoto Protocol, to which governments are being asked to sign on to. The Kyoto supporters argue that the dissenters are not only a *minority* but, by innuendo, are also less reputable. As history can attest—think of Galileo and Einstein!—scientific truth is not found by counting the numbers of adherents to and sceptics of a scientific hypothesis.[31] Even in the scientific community there is a phenomenon called 'the bandwagon effect'.

The bandwagon effect is dramatically illustrated by the large shift in the consensus on the climate change issue: in the 1970s, many eminent scientists declared a belief that the most imminent threat was posed by global cooling.[32]

[30] For a full discussion on this theme, see Siddayad (1993).

[31] The 'bandwagon' or 'political correctness' syndrome should not be discounted. As evidence of its importance in this context it is instructive to note the somersault of reputable scientists. See the column by Charles Krauthammer in www.globalwarming.org/Kraut.htm.

[32] To take but three examples cited by Charles Krauthammer in the same column (www.washingtonpost.com): a famous scientist, Nigel Calder, a

This zigzagging process within the span of a generation can be devastating for the public acceptance of the scientific process and their pronouncements. Should the public be moved to act on the basis of exaggerated fears and exaggerated faith in the measures put forward to stave off the dreaded denouement—and at some stage this is seen by the public to be manipulative hype—there would likely be a reaction that would reduce support for *effective* policies and programmes on the much wider range of environmental issues. It is important, therefore, to be on guard against evidence in the interpretation of findings of bias in selection of what is being publicized or by bias in interpretation. There is disturbing evidence of this bias—or, to be generous, sloppiness or laziness—when governmental and non-governmental organizations and media reporters and commentators cite sources such as reports of the UN's Intergovernmental Panel on Climate Change and declare that there is a 'scientific consensus' about future climate trends and the role of human activity in this phenomenon as though all scientists agree on both of those issues. But this is compounded when there is little or no publicity given to the third latest report (IPCC 2001), that states in its science section,

In sum, a strategy must recognize what is possible. In climate research and modeling we should recognize that we are dealing with a coupled non-linear system, and therefore that the prediction of a specific future climate is not possible.

---

former editor of *New Scientist*, wrote in 1975 that 'the threat of a new ice age must now stand alongside nuclear war as a likely source of wholesale death and misery for mankind; the editor of *Science Digest* urged during the same period that the atmospheric pollution needs to be carefully monitored as this 'will have a direct bearing on the arrival and nature of this weather crisis, i.e., a new ice age'; J. Murray Smith of the National Oceanic and Atmospheric Administration in 1976 observed that 'whenever there is a cold wave, the media seek out a proponent of the ice-age-is-coming school and put his theories on page one . . . . Whenever there is a heat wave . . . they turn to his opposite number for a prediction of a kind of heat death of the earth.'

The intensity of the controversy regarding truth and falsehood and the use and abuse of modelling is reminiscent of the debates revolving around the Malthusian hypothesis of the early 1970s and the preceding two centuries when William Pitt and the political elite played with the public of his day in using the Malthusian hypothesis and as later day political leaders used variants of this hypothesis in defence of the prevailing economic systems. When it is seen as a form of 'con-game'—as it unravels from its simplified version—the loss of credibility will be a great setback for the cause of achieving a manner of development that places a high priority on the future.

Nonetheless Should the Use of the 'Sustainable Development' Logo be Continued?

There are no failsafe, simple solutions to the growing propensity of decision-makers to embrace counter-productive intellectual fads; we need to be alert to our exposure to bad yet seductive ideas . . . (at a time when) globalization and cheaper instant electronic communication allow bad ideas to spread faster, . . . (when) the exponential growth of the 'noise' within our system of communications makes it harder to differentiate the bad ideas from the better ones . . . (and when) bad ideas now stand a better chance of becoming accepted thanks to the accelerated decision-making cycles and the increased public pressures that the decision-makers face.

Moises Naim, 'Misguided Ideas in a Dangerous World', *Foreign Policy*,
25 November 2002.

This cautionary note by the editor of the journal, *Foreign Policy*, was applied with special relevance to the policies related to the terrorism threat but applies equally well to the issue of the development/environment relationship. Just as new ideas are not necessarily better ideas, new terminology is not necessarily better than the old though it may sell better. It is time to retire the 'sustainable development' logo, a process that would take some time as the terminology is well and deeply entrenched. The best that can be done would be to discourage the vacuous language of this particular bad idea that obfuscates, deflects and terrorizes—and thus leads to bad policies for, as *The Economist* rightly notes in its 6 July 2003 issue, environment policy-makers are 'flying blind'.

In the same issue of *The Economist*, a proposal with regard to environment policy was put forward[33], namely, 'encourage science and technology, especially innovations that reduce the ecological footprint of consumption'. Leaving aside the issue of what *The Economist* means by the 'ecological footprint of consumption'— which is to put to one side the dynamic motor of the market economy—the kernel of the proposal is a very substantial increase in expenditure and by all other means to accelerate environmentally-related research on the requisite scale and with the requisite speed.

There are two precedents that are global in scope and could serve as examples of what could be done that would go a long way to help address the challenge of maintaining global growth with greater equity and, at the same time, reduce the stress of that growth on the environment. The first precedent is the Manhattan Project of the early 1940s. This was a research undertaking that mobilized talent *with no heed to cost* for the purpose of developing an atom bomb as a means of meeting a possible threat from the Nazis who, it was believed, were working on the manufacture of such a weapon. The immensity of the possible tragedy called for speed and an unqualified commitment of money and talent, sufficient to forestall such a tragedy.

The second precedent is the establishment, in 1970, of the internationally-funded entity called the Consultative Group on International Agricultural Research (CGIAR). Through its research programme, it has proven exceptionally successful since it was established over three decades ago in helping global food production to triple while world population doubled.

If an ambitious research programme is to serve as one of the key initiatives to address many of the facets of the environmental challenge of our times, it needs to be organized on a scale that

---

[33] These are discussed in Miller (1997).

is commensurate with the global dimensions of the challenge in its size and diversity. The governmental role is, therefore, critical. The participation of the private sector, including foundations and civil society, in the form of non-governmental organizations that constitute the 'sustainable development movement', could undoubtedly play a role in this endeavour, but for various reasons cannot be expected to lead. The outstanding example of that participatory role is the action taken by three private foundations, the Rockefeller, Ford, and Kellogg Foundations, that originated the idea of establishing the CGIAR and approached the World Bank to assume the role of leader. A similar approach could be attempted in other sectors such as environmentally-benign energy (*à la* CGIE2R?),[34] potable water,[35] and, as well, education, especially primary education in rural areas.[36]

All of these initiatives could lay the basis in time for a more productive population that, in turn, would be better able to leave a more prosperous, equitable, and environmentally healthy planet which is the most worthwhile legacy to future generations. Since the environmental issue has now captured the world's attention there is an opportunity for effective action, which would likely be enhanced by abandoning the 'sustainable development' concept that, by virtue of its ambiguity of meaning and its operational vacuity, is seriously flawed.

---

[34] E standing for energy and environment. For an elaboration see, 'High-tech to the Rescue? The Role H-T could play involving Rural People in the Knowledge Economy', University of Ottawa, School of Management, Working Paper No. 97–25, ISSN 0701-3086, 1999, in www.management.uottawa.ca/miller

[35] Ismail Serageldin, in one of his speeches when he was a Vice-President of the World Bank, called attention to the fact that more than a billion people have no access to clean water, that a child dies every second from drinking contaminated water, and that by 2025 the number will likely grow to more than three per second.

[36] These proposals are elaborated in papers to be found in www.management.uottawa.ca/miller. See especially op. cit., 'High-tech to the Rescue? . . .'

Interestingly enough, this assessment accords with that of the *The Economist*, when, in its 6 July 2002 issue, the journal's editors reversed themselves on the matter of the merit of 'sustainable development' *à la* Brundtland as a paradigm for policy on global environmental issues. In referring to 'Rio's fatal flaw' and to the fact that 'by nearly universal agreement, those grand aspirations have fallen flat in the decade since that summit', authors of the editorial acknowledged the mistake of subsuming the many facets of environment policy under the rubric of the concept called 'sustainable development':

The main explanation for the disappointment—and the chief lesson for those about to gather in South Africa (for the Johannesburg Conference on Sustainable Development)—is that . . . (at Rio) its participants were so anxious to reach a political consensus that they agreed to the Brundtland definition of sustainable development. . . . The biggest mistake is that (there has been an inclination to) slide over the difficult trade-offs between environment and development in the real world. To insist that the two are 'impossible to separate', as the Brundtland commission claimed, is nonsense.

Nothing better?!

References

Ackerman, F., 2001, 'Material Use and Sustainable Affluence', in J. Harris, T. Wise, K. P. Gallagher, and N. Goodwin (eds), *A Survey of Sustainable Development: Social and Economic Dimensions*, Island Press, Washington, D.C.

Agarwal, A., and S. Narain, 2001, 'Global Warming in an Unequal World: A Case of Environmental Colonialism', in J. Harris, T. Wise, K. P. Gallagher, and N. Goodwin (eds), *A Survey of Sustainable Development: Social and Economic Dimensions*, Island Press, Washington, D.C.

Alperovitz, G., 2000, 'Sustainability and Systemic Issues in a New Era', in J. M., Harris (ed.), *Rethinking Sustainability: Power, Knowledge and Institutions*, University of Michigan Press, Ann Arbor.

Bailey, R., 1993, *Ecoscam: The False Prophets of Ecological Apocalypse*, St Martin's Press, New York and London.

Banuri, T., G. Hyden, C. Juma, and M. Rivera, 1994, 'Sustainable Human Development: From Concept to Operation—A Guide for the Practitioner', UNDP Discussion Paper, UNDP, New York.

Barkin, D., 2000, 'Wealth, Poverty and Sustainable Development', in J. M. Harris (ed.), *Rethinking Sustainability: Power, Knowledge and Institutions*, University of Michigan Press, Ann Arbor.

Barrett, S., 1999, 'Montreal versus Kyoto: International Cooperation & the Global Environment', in I Kaul, I Grunberg, and M Stern, *Global Public Goods: International Cooperation in the 21st Century*, Oxford University Press, New York and London, pp. 192–19.

Beckerman, W., 1992, Economic Growth and the Environment: Whose Growth? Whose Environment? *World Development I.*

Benton, T., 1989, 'Marxism and Natural Limits', *New Left Review*, no. 178, November–December.

Biswas, A.K., M. Biswas and K. Tolba (eds), 1991, *Earth and Us: Population-Resources-Environment-Development*, UNEP, Butterworth-Heinemann, Oxford.

Bojo, J. and C. Reddy, 2002 'Poverty Reduction Strategies and Environment: A Review of 40 Interim and Full PRSPs', World Bank Environment Department Paper No. 86, Washington, D.C.

Brown, L., 1978, The 29th day: Accomodating Human Needs and Numbers to the Earth Resources, World Water Institute, Washington, D.C.

Bruyn, S., 2001, 'Civil Associations Toward a Global Civil Economy', in J. Harris, T. Wise, K.P. Gallagher, and N. Goodwin (eds), *A Survey of Sustainable Development: Social and Economic Dimensions*, Island Press, Washington, D.C.

Common, M., 1995, *Sustainability and Policy*, Cambridge University Press, Cambridge.

Commonwealth Secretariat, 1991, *Sustainable Development: An Imperative for Environmental Protection: Report by a Group of Experts on Environmental Concerns and the Commonwealth*, London.

Corcoran, T., 1990, 'Sustainable Development: A Dumpsite for Ideas', *The Globe and Mail* (Toronto), 23 March.

Costanza R. and H. Daly, 2001, 'Natural Capital and Sustainable Development', in J. Harris, T. Wise, K.P. Gallagher, and N. Goodwin (eds), *A Survey of Sustainable Development: Social and Economic Dimensions*, Island Press, Washington, D.C.

Court, T. de la, 1990, *Beyond Brundtland: Green Development in the 1990s*, Zed Books, London.

Daly, H. and J.B. Cobb, 1989, *For the Common Good*, Beacon Press, Boston.

———, 1990, 'Sustainable Growth: An Impossibility Theorem', *Development: Journal of Society for International Development*, (*Special Issue on Environment & Global Sustainability*), Rome.

Dasgupta, S., B. Laplante, H. Wang and D. Wheeler, 2002, 'Confronting the Environmental Kuznets Curve', *The Journal of Economic Perspectives*, American Economic Association, Winter, vol. 16, no. 1, pp. 147–68.

Easterlin, R., 1981, 'Why isn't the Whole World Developed?', *Journal of Economic History*, March.

El Serafy, S., 2001, 'Green Accounting and Economic Policy', in J. Harris, T. Wise, K.P. Gallagher, and N. Goodwin (eds), *A Survey of Sustainable Development: Social and Economic Dimensions*, Island Press, Washington, D.C.

Essex, C. and R. Mc Kitrick, 2002, 'Taken by Storm: The Troubled Science, Policy and Politics of Global Warming', Key Portes Books, Toronto.

Foster, J. B., 1995, 'Marx and the Environment', *Monthly Review: An Independent Socialist Magazine*, New York, vol. 47, no. 3, July/August.

———, 1998, 'Malthus' Essay on Population at Age 200: A Marxian View', *Monthly Review: An Independent Socialist Magazine*, New York, vol. 50, no. 7, December.

———, 2001, 'Ecology Against Capitalism', *Monthly Review: An Independent Socialist Magazine*, New York, vol. 53, no. 5, October.

———, 2002, 'Capitalism and Ecology: The Nature of the Contradiction', *Monthly Review: An Independent Socialist Magazine*, New York, vol. 54, no. 4, September.

Fredriksson, P. G., 1999, 'Trade, Global Policy and the Environment', World Bank Discussion Paper No. 402, Washington, D.C.

Freeman, M. A., 2002, 'Environmental Policy Since Earth Day 1: What Have We Learnt?', *The Journal of Economic Perspectives*, American Economic Association, Winter.

French, H., 1995, *Partnership for the Planet: An Environmental Agenda for the U.N.*, Worldwatch Institute, Washington, D.C.

———, 2000, 'Coping with Ecological Globalization', *State of the World 2000*, Worldwatch Institute, Washington, D.C.

Fulford, R., 2002, 'Giving Cliche's Short Shrift: They are a Crime Against Language', *National Post* (Toronto), July 9.

Gallagher, K., 2001, 'Overview Essay: Globalization and Sustainability', in J. Harris, T. Wise, K.P. Gallagher, and N. Goodwin (eds), *A Survey of Sustainable Development: Social and Economic Dimensions*, Island Press, Washington, D.C.

Hansen, S. et al., 1990, *Economic Policies for Sustainable Development: Report Synthesizing Seven Country Studies*, Asian Development Bank, Manila.

Harris, J. 2000a, 'Introduction: An Assessment of Sustainable Development', in J.M. Harris (ed.), *Rethinking Sustainability: Power, Knowledge and Institutions*, University of Michigan Press, Ann Arbor.

————, 2000b, 'Free Trade or Sustainable Trade? An Ecological Economics Perspective', in J.M. Harris (ed.), *Rethinking Sustainability: Power, Knowledge and Institutions*, University of Michigan Press, Ann Arbor.

————, 2001, 'Overview Essay: Economics of Sustainability: The Environmental Dimension', in J. Harris, T. Wise, K.P. Gallagher, and N. Goodwin (eds), *A Survey of Sustainable Development: Social and Economic Dimensions*, Island Press, Washington, D.C.

Harris, J. and N. Goodwin, 2001, 'Volume Overview', in J. Harris, T. Wise, K.P. Gallagher, and N. Goodwin (eds), *A Survey of Sustainable Development: Social and Economic Dimensions*, Island Press, Washington, D.C.

Hartwick, J., 1977, 'Intergenerational Equity and the Investing of Rents from Exhaustible Resources', *American Economic Review*, vol. 67, pp. 972–4.

Heal, G., 1999, 'New Strategies for the Provision of Global Public Goods: Learning from International Environmental Challenges', in I. Kaul, I. Grunberg, and M. Stern (eds), *Global Common Goods: International Cooperation in the 21st Century*, Oxford University Press, New York and London.

Henderson, H., 1996, *Creating Alternative Futures: The End of Economics*, Putnam, New York.

Holling, C. S., 2001, 'An Ecologist View of the Malthusian Conflict', in J. Harris, T. Wise, K.P. Gallagher, and N. Goodwin (eds), *A Survey of Sustainable Development: Social and Economic Dimensions*, Island Press, Washington, D.C.

Illich, I., 1974, *Energy & Equity*, Harper & Row, New York.

IPCC, 2001, Climate Change: Synthesis Report/Summary for Policy Markers, Intergovernmental Panel for Climate Change, Third Assessment Report, WMO, UNEP.

Jha, R. and J. Whalley, 1999, 'The Environmental Regime in Developing Countries', National Bureau of Economic Research Working Paper No. 7305, Cambridge, Massachusetts.

Johnson, B. and F. Duchin, 2000, 'The Case for the Global Commons', in J.M. Harris (ed.), *Rethinking Sustainability: Power, Knowledge and Institutions*, University of Michigan Press, Ann Arbor.

Kjorven, O. and H., Lindhjem, 2002, 'Strategic Environmental Assessment in World Bank Operations: Experience to Date—Future Potential', World Bank Strategy Note No. 4, Washington, D.C.

Kneese, A. V., 1977, *Economics and the Environment*, Penguin Books, New York and London.

Lange, G. M., 2002, '*Policy Applications of Environmental Accounting*', World Bank Environment Department Paper No. 87, Washington, D.C.

Lind, R. C. and R.E. Schuler, 2001, 'Equity and Discounting in Climate-Change Decisions', in J. Harris, T. Wise, K.P. Gallagher, and N. Goodwin (eds), *A Survey of Sustainable Development: Social and Economic Dimensions*, Island Press, Washington, D.C.

Linde, C. van der and M. Porter, 2001, 'Toward a New Conception of the Environment–Competitive Relationship', in J. Harris, T. Wise, K.P. Gallagher, and N. Goodwin (eds), *A Survey of Sustainable Development: Social and Economic Dimensions*, Island Press, Washington, D.C.

Lipietz, A., 2001, 'Enclosing the Global Commons: Global Environmental Negotiations in a North–South Conflictual Approach', in J. Harris, T. Wise, K.P. Gallagher, and N. Goodwin (eds), *A Survey of Sustainable Development: Social and Economic Dimensions*, Island Press, Washington, D.C.

Lovei, M., and B.S. Gentry, 2002, 'The Environmental Implications of Privatization: Lessons for Developing Countries', World Bank Discussion Paper No. 426, Washington, D.C.

Lvovsky, K., 2002, 'Environment, Health and Poverty', World Bank Strategy Note No. 1, Washington, D.C.

Magdoff, F., 2002, 'Capitalism Twin Crises: Economic and Environmental', *Monthly Review: An Independent Socialist Magazine*, New York, vol. 54, no. 4, September.

Matthews, W. H., 1980, 'Moving Beyond the Environmental Rhetoric', *Mazingira: The International Journal for Environment and Development*, vol. 4, no. 2.

McKibbin, W. and P. Wilcox, 2002, 'The Role of Economics in Climate Change Policy', *The Journal of Economic Perspectives*, American Economic Association, Spring.

MacNeill, J., P.Winsemius and T. Yakushiji, 1991, *Beyond Interdependence: The Meshing of the World's Economy and the Earth's Ecology*, Oxford University Press, New York and London.

Mészáros, I., 2001, 'Sustainable Development and Equality', *Monthly Review: An Independent Socialist Magazine*, New York, vol. 53, no. 7.

M. Miller, (ed.) 1962, *Resources for Tomorrow*, 3 volumes, Papers of Conference of Federal and Provincial Governments Queen's Printer, Ottawa.

———, 1983, 'The Challenge of the Energy Transition: the UN Response', in E. El-Hinnawi, M. Biswas and Asit K. Biswas, *New and Renewable Sources of Energy*, Volume 14 of Natural Resources and Environment Series, Tycooly International Publishing, Dublin.

———, 1990, 'Can Development be Sustainable?', *Development: Journal of Society for International Development*, (Special Issue: Environment and Global Sustainability), Rome, Italy.

————, 1993, 'Sustainability and the Energy/Environment Connection: Overcoming the Institutional Obstacles to "Doing the Right Thing"', in Corazon Siddayao (ed.), *Investing in Energy and the Environment*, World Bank, Washington, D.C., Chapter 4.

————, 1994, 'Promoting Bio-energy and the Environment: The Role of Large-Scale UN Conferencing', University of Ottawa Working Paper No. 94–57 (ISSN 0701–3086), Ottawa (in www.management.uottawa. ca.miller).

————, 1995, 'The Environment Policy Challenge', *Development Policy*, in S. Sharma (ed), St. Martin's Press, New York and London, Chapter 8.

————, 1996, *Debt and the Environment: Converging Crises?*, United Nations Publications, New York.

————, 1997, *High Technology to the Rescue?: The Role H–T Could Play Involving the Rural Poor in the Knowledge Economy*, University of Ottawa Working Paper No. 97–25, July (ISSN–0701–3086) or www.manage ment.uottawa.ca.miller.

————, 1998, 'The Chicken-little Syndrome and the Responsibility of Social Scientists', University of Ottawa Working Paper, School of Management, Ottawa.

————, 1999, 'Decentralized Energy: An Approach to Meet the Needs of the Rural Poor on a Scale Commensurate with the Challenge', in E. Bietry and V. Mubayi, *Decentralized Energy Alternatives*, Columbia University, New York.

Mishan, E.J., 1969, *The Costs of Economic Growth*, Pelican Books, London.

————, 1969a, 'Welfare Economics: Ten Introductory Essays', Random House, New York.

————, 1969b, 'Welfare Economics: An Assessment', North-Holland, Amsterdam.

————, 1993, 'Economic Growth: The Need for Skepticism', Lloyd's Bank Review.

Munasinghe, M. (ed.), 1993a, 'Environmental Economics and Sustainable Development', World Bank Environment Paper No. 3, Washington, D.C.

————, 1993b, 'Environmental Economics and Valuation in Development Decisionmaking', in M. Munasinghe (ed.), *Environmental Economics and Natural Resource Management in Developing Countries*, Committee of International Development Institutions on the Environment (CIDIE), World Bank, Washington, D.C.

————, 1993c, 'Issues and Options in Implementing the Montreal Protocol in Developing Countries', Munasinghe (ed.), *Environmental Economics and Natural Resource Management in Developing Countries*, Committee of International Development Institutions on the Environment (CIDIE), World Bank, Washington, D.C., Chapter 11.

Munda, G., 2001, 'Environmental Economic, Ecological Economics and the Concept of Sustainable Development', in J. Harris, T. Wise, K.P. Gallagher, and N. Goodwin (eds), *A Survey of Sustainable Development: Social and Economic Dimensions*, Island Press, Washington D.C..

Nordhaus, W. D., 1990, 'Greenhouse Economics: Count Before You Leap', *The Economist*, 7 July.

Oakes, J.B., 1989, Bush's Shell Game (editorial), *New York Times*, 12 January.

Pearce, D., 1989, 'Sustainable Development: An Overview', *Development: Journal of the Society for International Development*, (*Special Issue: Sustainable Development: From Theory to Practice*), vol. 2, no. 3.

———, 1996, 'Towards a Sustainable World', *Scientific American*, (*Special Issue: Managing the Earth*), New York, vol. 3, no. 4.

Pearson, C. S., 1985, 'Down to Business: Multinational Corporations, Environment and Development', Study No. 2, World Resources Institute, Washington, D.C.

Pezzey, J., 1992, 'Sustainable Development Concepts: An Economic Analysis', World Bank Environment Paper No. 2, Washington, D.C.

Pezzey, I., 1992, 'Economic Analysis of Sustainable Growth and Sustainable Development, Environment Working Paper No. 15, World Bank, Washington.

Redclift, M., 1987, *Sustainable Development: Exploring the Contradictions*, Methuen, London and New York.

Reddy, A., R.H. Williams and T.B. Johansson, 1997, *Energy After Rio: Prospects and Challenges*, United Nations Development Programme (UNDP), New York.

Reed, D., 2001, 'Impacts of Structural Adjustment on the Sustainability of Developing Countries', in J. Harris, T. Wise, K.P. Gallagher, and N. Goodwin (eds), *A Survey of Sustainable Development: Social and Economic Dimensions*, Island Press, Washington, D.C.

Repetto, R., 1989, 'Wasting Resources: Natural Resources in the National Income Accounts', World Resources Institute, Washington, D.C., June.

Repetto R. and D. Austin, 2001, 'The Costs of Climate Protection: A Guide for the Perplexed', in J. Harris, T. Wise, K.P. Gallagher, and N. Goodwin (eds), *A Survey of Sustainable Development: Social and Economic Dimensions*, Island Press, Washington, D.C.

Ross, E. B., 1998, 'Malthusianism, Counter-revolution and the Green Revolution', *Organization & Environment*, vol. 12, no. 1, December.

Roumasset, J., 1990, 'Economic Policy for Sustainable Development', *Development: Journal of Society of International Development, (Special Issue: Human-Centred Economics)*, vol. 3, no. 4, Rome.

Ruckelhaus, W., 1989, 'Towards a Sustainable World', *Scientific American, (Special Issue: Managing Planet Earth)*, September.

Ryle, M., 1988, *Ecology and Socialism*, Radius, London.

Sachs, I., 1997, 'Rio, Five Years Later: Against a Wintry Sky, a Few Swallows', *Ecodecision: Environment and Policy Magazine*, IRPP, Montreal, Spring.

————, 1980, *Strategies de l'Ecodeveloppement*, Les Editions Ouvrières, Paris.

Sachs, W., 2001, 'Global Ecology and the Shadow of Development', in J. Harris, T. Wise, K.P. Gallagher, and N. Goodwin (eds), *A Survey of Sustainable Development: Social and Economic Dimensions*, Island Press, Washington, D.C.

Schmalensee, R., P.L. Joskow, A.D. Ellerman, J.P. Montero and E.M. Bailey, 1998, 'An Interim Evaluation of Sulfur Dioxide Emission Trading', *The Journal of Economic Perspectives*, American Economic Association, Summer.

Shiva, V., 2001, 'Conflicts of Global Ecology: Environmental Activism in Period of Global Reach', in J. Harris, T. Wise, K.P. Gallagher, and N. Goodwin (eds), *A Survey of Sustainable Development: Social and Economic Dimensions*, Island Press, Washington, D.C.

Siddayad, C. (ed.), 'Sustainability and the Energy/Environment Connection: Overcoming the Institutional Obstacles to Doing the Right Thing'.

Solow, R., 1974a, 'The Economics of Resources or the Resources of Economics', *American Economic Review.*

————, 1974b, 'Intergenerational Equity and Exhaustible Resources', *Review of Economic Studies.*

————, 1991, 'Sustainability: An Economist's Perspective', *Journal of the Woods Hole Oceanographic Institution.*

————, 1992, 'An Almost Practical Step Towards Sustainability', *Resources for the Future*, Washington, D.C.

Stavins, R. N., 1998, 'What Can We Learn from the Grand Policy Experiment? Lessons from SO$_2$ Allowance Trading', *The Journal of Economic Perspectives*, American Economic Association, Summer.

Stiglitz, J., 1998, 'The Private Use of Public Interests: Incentives and Institutions', *The Journal of Economic Perspectives*, American Economic Association, Spring.

————, 1999, 'Knowledge as a Global Public Good', in I. Kaul, I. Grunberg, and M. Stern (eds), *Global Common Goods: International Cooperation in the 21st Century*, Oxford University Press, New York and London.

Stone, C. D., 1993, *The Gnat is Older than Man: Global Environment and Human Agenda*, Princeton University Press, Princeton, New Jersey.

Stott, P., 2000, 'Political Ecology: Science, Myth and Power', Oxford University Press, New York and London.

Stretton, D., 1988, *Capitalism, Socialism and Environment*, Cambridge University Press, Cambridge.

Strong, M., 2001, 'More is Not Enough', *Scientific American* (Special Issue: The Growth-Environment Dilemma), vol. 3, no. 4.

Sudhir, A., and A. K. Sen, 1996, 'Sustainable Human Development: Concepts and Priorities', UNDP Discussion Paper, UNDP, New York.

Tellus Institute, *Halfway to the Future: Reflections on the Global Condition*, Boston, Mass.

The Economist, September 1989; September 1992; December 1997; July 2002; July 2003.

The Economist, 1991, 'Energy and the Environment', (Special issue), 31 August.

The Globe and Mail (Toronto) 2002, 'Don't blow it Canada', December 6, *The Time*, 2002 July 1.

Thorsell, W., 2002 'Rising it All', *The Globe and Mail*, June 10.

Tietenberg, T., 1990., 'Using Economic Incentives to Maintain our Environment', *Challenge: The Magazine of Economic Affairs*, March–April.

Toman, M. A., 2001, 'Economics and "Sustainability": Balancing Trade-offs and Imperatives', in J. Harris, T. Wise, K.P. Gallagher, and N. Goodwin (eds), *A Survey of Sustainable Development: Social and Economic Dimensions*, Island Press, Washington, D.C.

Tucker, W., 1980, 'Environmentalism: The Newest Toryism', *Policy Review*, no. 14, Fall.

Vylder, S. de, 1995, 'Sustainable Human Development and Macroeconomics: Strategic Links and Implications', UNDP Discussion Paper, UNDP, New York.

Watson, R.T., J.A. Dixon, S.P. Hamburg, A.C. Janetos and R.H. Moss, 1998, *Protecting Our Planet: Securing Our Future*, UNEP, NASA, and the World Bank, Washington, D.C.

Wente, M., 2002 'The Kyoto-Speak Brainwashes', *The Globe & Mail (Toronto)*, 7 December.

White, A.L., 2001, 'Sustainability and the Accountable Corporation', in J. Harris, T. Wise, K.P. Gallagher, and N. Goodwin, (eds) *A Survey of Sustainable Development: Social and Economic Dimensions*, Island Press, Washington, D.C.

World Bank (Annuals), *Environment Matters at the World Bank: Towards Environmentally and Socially Sustainable Development*, Washington, D.C.

————, 2002a, *The Environment and the Millenium Development Goals*, Washington, D.C.

————, 2002b, *Third Environmental Assessment Review (FY1996–2000)*, Washington, D.C.

World Commission on Environment and Development (Brundtland Commission), 1987, *Our Common Future*, Oxford University Press, New York and London.

World Wildlife Fund, 2002, Living Planet Report, www.panda.org

Young, O., 1989, *International Cooperation Building Regimes for Natural Resources and the Environment*, Cornell University Press, Ithaca, New York.

3

Sustainable Development: Some Unanswered Questions

Asit K. Biswas

Introduction

At the dawn of the twenty-first century, any objective and in-depth analysis of the total long-term impacts of the official development assistance will indicate that these have generally had at best only a marginal impact in alleviating poverty, improving the quality of life of billions of people, and maintaining and/or improving the conditions of the natural environment and the ecosystems.

During the past three decades, the international system has consistently made numerous commitments and pledges that were expected to alleviate poverty very substantially, or even eradicate it completely. For example, at the World Food Conference, convened by the United Nations in Rome in 1974, senior decision-makers from all parts of the world, at the explicit recommendation of the former Secretary of State of the United States, Henry Kissinger, made a pledge that within a decade no child anywhere in the world would go to bed hungry. More than a quarter of a century has elapsed since the world leaders and the United Nations made that commitment, but children continue to go to bed hungry, perhaps even in larger numbers than before. The situation continues to be as grim as ever. In some aspects, and in many parts of the world, the conditions have even deteriorated significantly.

Similarly, the two Development Decades initiated by the United Nations System in the 1970s and 1980s also had somewhat marginal impacts. In fact, the impacts of the First Development Decade was so minimal that Bradford Morse, a former Administrator of the United Nations Development Programme, an institution that was made officially responsible for implementing the two Development Decades, called it formally 'the lost decade' during his own terms of office, when he was responsible for administering the Second Development Decade. While such candour for truth is refreshing and somewhat unusual for a senior international bureaucrat, the fact still remains that the world has made only limited progress in terms of eradicating poverty and improving the quality of life for more than one billion human beings.

Inspite of the commitments made by the global leaders, and the continued rhetoric of the international institutions, poverty and hunger have continued to be as pervasive as ever, and have even increased significantly in recent years in many parts of the world. Furthermore, the gulf between the rich and the poor, both between countries and within countries, has increased, rather than decreased, in recent decades. Similarly, the environmental conditions have continued to deteriorate in most parts of the world.

Development Goals and Their Achievement

Consider the following statistics released by the World Bank (2001) on the current development conditions of the world, which by any national or international standard would be considered unacceptable.

- Out of the current global population of 6 billion, 2.8 billion (47 per cent) live on an income of less than $2 per day, and 1.2 billion (25 per cent) live on less than $1.00 per day. Some 44 per cent of the world's absolute poor (daily income less than $1.00) live in South Asia.

- In South Asia, sub-Saharan Africa, and Latin America, the total numbers of poor people have been rising steadily. For the countries in transition in Eastern Europe and Central Asia, the number of absolute poor (income less than $1.00 per day) has risen more than 20 times in recent years.
- In poor countries, 50 per cent of all children under five years of age and 5 per cent in rich countries, are currently malnourished.
- The average income in the richest 20 countries of the world is 37 times the average income of the poorest 20 countries. This income gap has not decreased in recent decades, in fact, the gap has actually doubled during the past 40 years.

Faced with this dismal global performance in terms of alleviating poverty in the developing countries during the past three decades, various United Nations Conferences during the 1990s, at high decision-making levels, have reformulated international development goals in terms of reduction of poverty and human deprivations. These objectives of international development were also separately agreed to by the countries belonging to the Organization for Economic Co-operation and Development (OECD), that is, the developed world. These new development goals include, *inter alia*, the achievement of the following by the year 2015 (OECD, 2002):

- reducing by half the proportion of people living in extreme poverty (income less than $1.00 per day). This has to be achieved in a world whose population is estimated to increase by some 2 billion by 2025, with 97 per cent of this increase occurring in developing countries,
- ensuring universal primary education, and
- reducing by two-thirds infant and child mortality.

If the past attempts did not contribute to improvements in these indicators cost-effectively and within a reasonable time frame, as has clearly been the case, some fundamental questions need to be asked and answered: why have such attempts failed miserably and consistently? and what lessons can be learnt from such failures so that the future development policies do not make the same mistakes?

The main recommendation as to how these international development goals can be achieved, according to the latest high-level UN fora and the leaders of the OECD countries, would be through the implementation of national strategies for sustainable development in every country by 2015. This is inspite of the fact that the UN system as a whole has never defined what is meant by sustainable or unsustainable development in operational terms, nor identified the parameters that should be measured to indicate whether sustainable development is taking place or not. Nor has any single government or any international or national institution done so. No one has ever asked if sustainable development has been achieved in any country of the world, and if so, in which ones, how it was achieved, over what periods, and what have been the impacts on critical issues like poverty alleviation, income distribution, economic growth, overall quality of life, and environmental conservation. Also, no consideration has been given to conduct serious analyses to determine if the past attempts at sustainable development have actually improved the very same international development indicators, which the UN system have now identified as the goals to be achieved by 2015. Nor has the following fundamental question been asked: would the world have been any different now if it had not followed the currently fashionable concept of sustainable development over the past two decades?

It should be noted that the establishment of such arbitrary development goals are not new. It is seldom that good technical and economic analyses are made in order to realistically determine as to how these goals can be achieved, where will the additional funds for such development come from, and who will be responsible for formulating and implementing the development programmes. For example, for the water sector, both the United Nations Conference on Human Settlements (Vancouver, 1976) and the United Nations Water Conference (Mar del Plata, 1977) pledged that every human being should have access to clean water and sanitation by 1990. The UN General Assembly reconfirmed this goal by declaring the decade of the 1980s as the International Water Supply and Sanitation Decade. More than 10 years after the UN Decade is over, universal availability of clean water remains a goal that appears to be as illusive as ever. The situation in terms of sanitation is even worse.

Serious and objective evaluations and assessments of the reasons as to why the previous development targets were never reached are seldom undertaken. In contrast, there are many superficial and pseudo-evaluations, which have basically concluded that if more money was forthcoming, and if there was more 'political will', the problems would have been solved. No critical in-depth analyses are available to determine if the funds expended could have been used more efficiently, if the institutions concerned (both national and international) were competent to carry out the tasks they were entrusted with, if the right policies and competent and experienced personnel were in place, etc. Instead of considering and answering such difficult and complex questions, the simpler solution has been to declare partial victory, and then set a new target date to achieve the goals, which should have been achieved in the first place many years ago. The new target dates are set a decade or two later, to achieve the same objective as before. Thus, at least for the development target in the area of water supply and sanitation, the international community, having failed to meet the objective of universal water supply and sanitation by 1990, simply decided to extend the target date to a quarter of a century later, to 2015. On the basis of the latest trends, even this goal is highly unlikely to be achieved by the new target date. Thereupon, a cynic might say that the target date will be further postponed to another decade or two in the future.

Sustainable Development

There is no question that in the international political fora, sustainable development has become a powerful and all-embracing slogan during the past 15 years. Every government is for it, as are all the major international institutions such as the United Nations agencies, World Bank, regional development banks, and OECD, as well as all the environmental and social non-governmental organizations (NGOs). Major institutions have initiated programmes, or specific budget lines, for sustainable development. This is inspite of the fact that there is no agreement at present between the various parties concerned, as to what is meant by sustainable development, whether it works, and if so, under what conditions, what are its impacts (positive, negative, or neutral) on human lives and other appropriate development indicators, and how it can be achieved operationally in a real world, and especially in developing countries.

Contrary to popular belief, the concept of sustainable development is not new. The general philosophy behind the concept has been expounded for centuries, if not millennia. For example, William Shakespeare said in *Hamlet*:

Suit action to the word, the word to the action; with this special observance, that you overstep not the modesty of nature.

Similar thoughts on living in harmony with nature can be found in most religious texts.

The use of the term 'sustainable development' became fashionable around 1980. However, there is very little difference between this and other earlier concepts like ecodevelopment, basic human needs-outer limits, or environment and development that were prevalent during the 1970s. Neither of these concepts could be made operationally possible in a real world, and thus these paradigms slowly disappeared during the early 1980s, only to be replaced with another very similar one, i.e. sustainable development. In fact, one would be hard-pressed to conceptually differentiate between the earlier concept of ecodevelopment with the current paradigm of sustainable development.

Sustainability is unquestionably a popular concept at present, but it means different things to different people. One is reminded of the popular support for the Conservation Movement of the United States in the early part of the twentieth century, when President Theodore Roosevelt correctly said that 'Everyone is for conservation: no matter what it means!' The situation does not appear to be much different at present for sustainable development.

Sustainability: What is It?

The concept of sustainable development, as it is used at present, was basically borrowed from the field of fisheries management in the late 1970s, where it has been used successfully for well over half a century. However, in the case of fisheries, the concept is simple, measurable and implementable. It means that the amount of fish catch should be equal to or less than total reproduction so that fishery in any region can be sustainable ad infinitum.

Before sustainable development became fashionable, the term 'sustainability' was technically used for harvesting reproducible natural resources, e.g. maximum sustainable yield for fisheries. This concept was extended in the late 1970s by a group of environmental scientists meeting in Nairobi under the aegis of the United Nations Environment Programme. The broadened concept of sustainable development was expected to be a 'new' idea for assessing and managing human impacts on the environment and natural resources.

The term was later popularized by the Brundtland Commission report *Our Common Future*, which was published in 1982. The Commission defined sustainable development in a somewhat amorphous way as 'development that meets the needs of the present without compromising the ability of the future generations to meet their own needs'. Not surprisingly, with such a vague, simplistic, internally inconsistent, and static definition, the Commission was unable to specify what was to be sustained. The report made continual references to sustainability, but was totally unable to say how the concept could be operationalized. Sustainability was expected to be achieved in an unspecified and undetermined way, some time in the unspecified future. Nor did the definition include the realization of a reasonable and equitably distributed level of economic well-being, without which no development can be sustainable over the long term. This aspect is especially important for developing countries, where income distribution between the rich and poor has already become a socio-political issue.

Once the concept became popular, dozens of new definitions were offered. Currently one can easily identify more than one hundred definitions of sustainable development without much difficulty. The concept was promptly embraced by many institutions because of its simplicity and vagueness, which allowed them to define it in a way that best suited their interests and agenda. Thus, even though all the United Nations agencies now champion sustainable development, individual institutions often define it the way it is most convenient and beneficial to them. Thus, the definition of sustainable development often varies from one UN agency to another in some significant ways, even though all the UN agencies embraced this concept early two decades ago.

Sustainability: Some Major Issues

In spite of the present rhetoric, it has to be admitted that operationally it has not yet been possible to identify a development process which can be planned and then implemented in such a manner that it becomes inherently sustainable, however this may be defined. It would be true to say that there has been more success in identifying certain aspects of development which are unsustainable and then taking appropriate remedial steps to reduce or even eliminate those undesirable effects, compared to devising a holistic process that is intrinsically sustainable right from the very beginning.

For example, if sustainable water resources development is considered, it has been known for decades that irrigation without drainage would contribute to waterlogging and salinity, which in turn would reduce the yields of the irrigated area over a period of time. Since the main purpose of any irrigation project is to increase the total agricultural production, clearly any system that does not fulfil this objective over a long-term period cannot be considered to be sustainable. However, the provision of drainage alone will not make an agricultural system inherently sustainable.

There are many other factors, some tangible and others intangible, which, only when considered concurrently, are likely to define the sustainability of the system. Similarly, if extensive use of fertilizers by the farmers increases the nitrate content of groundwater so that its use for drinking purposes is impaired, then this practice has to be considered unsustainable. Again, there are numerous other factors, some known and others unknown, which contributed to the sustainability of using any groundwater system implementation.

While there are many issues that are important for sustainable development, from an implementation point of view, three factors need special consideration.

Short- versus Long-term Considerations

The concept of sustainable development automatically assumes that the process selected would continue over the long term, even though the issue of what constitutes 'long term' has neither been clarified nor featured much in the past or current discussions. The time factor, either inadvertently or because of its complexity, has basically been left fuzzy: no attempt has been made to define or even discuss what is meant by the long term. For example, does sustainability cover 50 years, or 100, 500, 1,000 years, or even more? Some have spoken vaguely of 'several' generations. Even if one considers the lowest figure of 50 years, there is a fundamental dichotomy as to its use in the real world.

Consider irrigated agriculture. Generally the economic planning horizon of the farmers extends to the next cropping season, or at most the next two such seasons. The overriding philosophy of nearly all the farmers anywhere in the world has been to maximize economic returns from their agricultural activities within this short and limited time frame. Thus, the mind set is inherently based on maximizing profits over a continual series of short-term periods without any specific or explicit considerations of their long-term benefits and costs. Though the short-term benefits could have long-term costs (e.g. in terms of soil erosion, salinity development, etc.), generally short-term considerations have won over the long-term implications. While in some cases this emphasis on short-terms could be due to a lack of knowledge or understanding of the potential long-term impacts of their activities, it has to be admitted that, for financial reasons, small farmers in developing countries, who are generally poor, are mostly forced to consider only the short-term economic implications for their own survival. Large farmers are no different in their perceptions and outlooks either.

Similarly, for large private sector companies, their performances are judged on the basis of their profits every 3 months, and their profit expectations over the next 4 to 8 quarters. The managers of these companies are rewarded on the basis of their quarterly and annual performances. The stock prices of the companies depend exclusively on their quarterly performances. While the politicians can get away with their promises of 'jam tomorrow', provided some sacrifices and hardships are faced in the short term, the private sector managers often will lose their jobs, unless they are capable of showing increasing quarterly profits, irrespective of the long-term impacts of their management practices. No manager will survive in the private sector if he/she is to promise the stakeholders that they will suffer for the next five–ten years in terms of low or no profit, but that their profit pictures thereafter will be magnificent. The market will clobber that company's stock mercilessly, and the President and CEO of that company will simply be fired by its own Board of Directors.

Accordingly, and inspite of the rhetoric of the World Business Council for Sustainable Development, no business can survive if only the long-term implications are considered, and short-term impacts are ignored.

Hence, even if the societal and/or governmental goal is to achieve long-term sustainable development, in reality the main objective of a vast majority of farmers and private sector companies often extends only to short-term benefits, which predominantly dictate their behaviours, perceptions, and approaches. Thus, any plan for sustainable development that does not specifically consider this fundamental conflict between short-term and long-term consideration and then attempts to identify realistic alternatives to overcome the problem, is doomed to fail. Such plans become primarily academic exercises which gather only dust on the shelves. The situation is very similar for private sector companies as well where short-term considerations often dictate long-term developments.

Externalities

Externalities occur when private costs or benefits do not equal social costs or benefits. People operate primarily on the basis of their own private costs and benefits. If they perceive opportunities which could reduce their costs and/or increase potential benefits, they often take actions which could be beneficial to them, even when they are unlikely to serve the common good. A common example is the discharge of wastes from municipalities and industrial concerns to rivers and other water bodies, which could impair existing water uses of numerous other people sharing the same water system. The private economic benefits due to non-treatment of wastewaters are likely to be significantly less than the societal costs of using polluted water.

Such costs could be internalized, at least conceptually, through taxes, subsidies, and regulations. But in reality, even in developed countries, it has not been possible to internalize the externalities for four important reasons. First, methodologically, calculation of the precise value of externalities has been a very difficult task. Often two experts may disagree in terms of their estimates of the external costs, and even the methods used to estimate them. Second, frequently there are politically powerful individuals and organizations who vociferously defend their own considerable private advantages against a large number of unorganized and disadvantaged individuals, or even the society as a whole, who may be experiencing additional costs, directly and/or indirectly. Third, externalities could develop steadily over time, and thus there could be a time gap before those affected realize the real costs, which over the years could become very substantial. Finally, regulations to control such externalities in nearly all developing countries have proved to be somewhat ineffective and expensive. Developed countries have had only marginally better success.

Risks and Uncertainties

A major issue confronting sustainable development relates to the risks and uncertainties that are inherently associated with any complex development process. For example, with the increasing population base of the Asian developing countries, there is no question that resources such as land and water have to be used intensively in order to maximize agricultural yields, and thus the total production. The fundamental questions, for which there are no real clear-cut answers at the present state of knowledge, are: up to what level can an agricultural production system be intensified, without sacrificing sustainability, irrespective of how it is defined? What early warnings could indicate the beginning of a transition process from a sustainable to an unsustainable system, or vice versa? What are the parameters that need to be monitored to indicate that such a transition is about to occur from one state to another, or, indeed, is occurring? Clearly, our present knowledge is inadequate even to identify all the parameters that could indicate the passage from one state to the other, and their relative primarily considerations. Thus, currently, it is not possible to accurately detect, much less predict, the transition of any sustainable system to an unsustainable one, or vice versa. In addition, all natural systems are variable. For example, a major global concern at present is climate change, especially in terms of changes in the existing precipitation and temperature regimes, both of which are subject to very wide natural variations. Their normal fluctuations could be so great that statistically significant data could be very expensive, or even impossible, to collect in order to state categorically that such variations are normal (that is within the existing standard deviations), or due to other reasons. When additional factors such as potential climatic changes are superimposed on inherently variable systems the degree of uncertainty in terms of detecting or predicting the transition process from one stage to another increases greatly. One is then confronted with the difficult issue of even identifying the direction of any change, let alone estimating the degree of change with any degree of reliability.

These types of fundamental issues need to be discussed and resolved successfully before the concepts of sustainable development can be holistically conceived and then implemented. Unfortunately, while much lip-service is given to sustainable development at present, most of the published works on this subject are either somewhat general, or a continuation of earlier 'business as usual' undertakings that have only been given the latest trendy label of 'sustainable development'. If sustainable development is to become a reality, national and international organizations will have to address many real and complex questions, which they have not done so far in any measurable and meaningful fashion; nor are there any signs that they are likely to do so in the foreseeable future. If not, and unless the current rhetoric can be translated effectively into operational reality, sustainable development will remain a trendy and fashionable paradigm for some years, and then gradually fade away like the earlier concept of ecodevelopment. It would then be replaced by a new and more fashionable paradigm.

It is indeed a curious irony that we have spent the last two decades discussing and promoting *what is not* sustainable development rather than what it is. We have concentrated almost exclusively on those aspects which cannot be sustained. By trying to define sustainable development in terms of only those factors that could contribute to unsustainability, clearly we have focused our entire attention only on one part of the equation, and have completely ignored the other, which could possibly be as important as the negative aspects, if not more so. Sustainable development, as it is analysed at present, focuses *only on what it is not*, and then attempts to ameliorate the potential negative effects. This issue is thus not approached holistically. Consideration should first be given to what is sustainable development, and then proceed to consider what is *un*sustainable. Instead, we are hung up exclusively on how to reduce the negative aspects of sustainable development. It is worth noting, even though it is axiomatic, that any significant development project would have many economic, social, and environmental impacts. However, the word 'impact' in the existing development context has primarily, and almost exclusively, negative connotations. While any large development project, irrespective of its nature, will have both positive and negative impacts, current analyses of environmental and social impacts generally consider *only* adverse impacts and their potential amelioration.

To a certain extent this overwhelming emphasis on the negative aspect of all major development projects can be explained by historical developments. During the 1970s and earlier, project analyses primarily consisted of technical and economic considerations: environmental and social issues were mostly not seriously analysed or considered. Because of this general neglect, and some very visible but adverse impacts of certain development projects on the society and the environment, a movement to promote environmental conservation gradually developed in the West. Within a very short period, environmental protection became an important item on the political agenda in the early 1970s in some developed countries, primarily through the activities of environmental pressure groups and NGOs.

Not surprisingly, this negative attitude and perception of environmental protection was reflected in the United Nations Conference on the Human Environment, held in Stockholm in June 1972. A retrospective analysis of the Stockholm Action Plan, as approved by all the UN member countries, clearly indicates its negative approach to environmental management: stop all pollution stemming from any development activity, stop exhausting non-renewable resources, and stop using renewable resources faster than their generation. The emphasis thus was primarily on controlling the adverse impacts of development—positive aspects did not receive much attention.

Accordingly, environmental impact analysis, which was developed and made mandatory in many developed countries during this era, was exclusively concerned with the identification and amelioration of negative impacts of development projects only; positive impacts were mostly ignored. Because of this inauspicious and incorrect beginning, the term 'impact' has continued to have almost exclusively negative connotations. Sadly, this unfortunate situation has not changed over the past two decades.

Concluding Remarks

It is clear that the development profession is facing a critical problem, of a magnitude and complexity not seen earlier. The development profession, now really has two stark choices: to carry on as before with a 'business as usual' attitude with only some marginally incremental changes, and thus endow our future generations with a legacy of suboptimal development process and management practices, or to continue in earnest an accelerated effort to identify and implement development processes that can successfully meet the expectations of humankind as a whole. Rhetoric and exhortations for sustainable development are no longer enough: paradigms and concepts must be implementable to solve global problems, both cost-effectively and within a reasonable time frame. The net result must be to improve the quality of life of the people of the world as a whole. Fashionable though the current paradigm of sustainable development may be, its usefulness, irrespective of its conceptual attraction and widespread acceptance, can only be marginal unless it can be used operationally and effectively in the real world. The concept of sustainable development must be critically appraised and reassessed. It may then be considered necessary to modify appropriately, or even jettison, the concept, unless it can be shown that it works in terms of its application to solve complex problems in the real world. We no longer have any soft options left, only hard choices. To quote George Bernard Shaw:

You see what is and ask, 'Why?'
I see what could be and ask 'Why not?'

References

OECD, 2002, 'The DAC Journal, Development and Cooperation, 2001 Report', OECD, Paris.

World Bank, 2001, 'World Development Report 2000/2001, Attacking Poverty,' World Bank, Washington, D.C.

4

Actors, Problem Perceptions, Strategies for Sustainable Development. Water Policy in Relation to Paradigms, Ideologies, and Institutions

Peter Söderbaum

Introduction

Humanity faces a number of environmental and development problems. Climate change, depletion of the ozone layer, loss of biodiversity, chemical pollution, and degradation of water quality are some examples on the environmental side. Individuals in some countries or at some places suffer more than others as a result of the mentioned and other development problems. Poverty is a major issue and the links between environmental issues and poverty are increasingly discussed (Department for International Development UK et al. 2002). Like welfare, poverty, or 'capability deprivation' (Sen 1999) can be understood in multidimensional terms and not reduced to monetary income. While some efforts have been successful in improving human conditions and the state of the environment, the tendency in other areas is rather that things are getting worse

*I have benefited from comments by Asit Biswas and Cecilia Tortajada, both from the Third World Centre for Water Management, Atizapan, Mexico, to an earlier draft of this paper.

(Commission of the European Communities 2001; UNEP 2002).

Why is this so? And what can be done about it? The first UN Conference on Environment and Development was arranged in 1972 in Stockholm and other such meetings have followed. Among UN initiatives that have focused on environmental issues in relation to development, the Rio de Janeiro Conference in 1992 was at least a partial success, with many national and international leading actors involved. In 2002, a 'Stockholm thirty-years on' was organized in an effort to contribute to the Johannesburg summit, held the same year.

In the case of water policy, when summarizing the activities and recommendations of the Stockholm Water Symposium over the years, it has been argued that there is still a need for 'major shifts in thinking' (Falkenmark 2000). The symposium of 2001 pointed to a similar direction (Tortajada et al. 2002; Water Science & Technology 2002). Another result of international co-operative activities is the Ministerial Declaration of a Conference on Freshwater in Bonn (International Conference on Freshwater 2001). Water was seen to be a 'Key to Sustainable Development' and a number of recommendations in terms of governance, capacity building, participation, protection of ecosystems, etc., were made.

Recommendations of the kind mentioned are all meaningful but some actors at the above mentioned Stockholm + 30 conference argued that what is needed is 'action' rather than 'words' or rhetoric (Ministry of the Environment 2002). They referred to a gap between what has been agreed and what has been implemented. Implementing some part of what has already been decided would then be a good strategy. While this way of reasoning should be taken seriously, some fundamental issues may not yet have been sufficiently addressed. Some words and arguments are still missing in the development dialogue. In this sense, one should try to identify areas and actors so far protected from serious dialogue and action. For such 'protected zones' there is a need for more rather than less words and analysis.

Neoclassical Economics as a Protected Zone in the Development Dialogue

As an example, establishment actors tend to avoid a serious debate about the role of economics as a theoretical perspective and paradigm in relation to development. However, mainstream neoclassical economics remains the most important theoretical perspective or paradigm influencing the mindsets of influential actors in present societies. A majority of establishment actors seem to believe that development and welfare can be reduced to economic growth in terms of gross domestic product (GDP), monetary profits for business companies, and so on. Not even the seriousness of present environmental and development problems has led to a questioning of the monopoly of neoclassical economics in most establishment circles. Instead, the tendency is to listen to neoclassical economists and their story about possible marginal failures. As is well known, neoclassical economists connect environmental problems with the possibilities of 'market failure' and 'government failure'. Market transactions may influence third parties negatively and in such cases 'externalities should be internalized' and the 'polluter pays principle' applied. Governments may subsidize activities that degrade natural resources and in such cases subsidies should be removed. The distinction between 'failure' and 'success' is built on the conventional neoclassical idea of efficiency as in Cost-Benefit Analysis, for instance.

Proposals to internalize externalities or remove subsidies with harmful impacts on the environment are all worthy of consideration and the problem is rather that they are too seldom applied in practice. This may in turn be explained by other more fundamental 'failures' in relation to some vision of a healthy development. Most of these potential failures are outside the scope of neoclassical analysis, for example paradigm failure; ideology failure; institutional failure; failure of organizations as actors; and failure of individuals as actors.

Something may be wrong with the kind of paradigms, ideologies, or institutional arrangements that have dominated for some time. 'Paradigm' here refers to conceptual and theoretical perspective, and neoclassical economics exemplifies a paradigm. Institutional theory is another paradigm with origin in economics but nowadays influential in many disciplines, such as economics, economic history, sociology, and business management. While neoclassical economics is useful for some purposes, the near monopoly position of neoclassical economics at university departments of economics in all parts of the world is a considerable problem and an example of 'paradigm failure'. Here pluralism is the key to 'success' for reasons that will be explained.

'Ideology' is used in a rather broad sense to refer to ideas about means and ends or 'means-ends philosophy'. 'Ideology' or 'ideological orientation' is therefore not limited to established political ideologies such as socialism or liberalism but includes various versions of 'ecologism' or 'Green' ideology. It is assumed that an individual is guided by her 'ideological orientation', i.e. patterns of thought and values, and that ideology therefore is not exclusively a collective phenomenon. Neoclassical economics, while being science in some sense, at the same time qualifies as a 'means-ends philosophy' and thereby ideology. In fact, neoclassical economics is more precise as ideology than most established political ideologies mentioned. Neoclassical theory recommends and even imposes a view of human beings as consumers at the expense of all other roles (citizens, professionals, parents, etc.), a view of organizations as firms or business companies at the expense of all other groups (churches, universities, civil society organizations), a specific view of markets in terms of supply and demand while other ideas of markets are not considered or play a negligible role, specific ideas of economics, efficiency, valuation, decision making, social change, and so on. Together these elements form not only a kind of microeconomics but at the same time a very specific ideology or ideological orientation.

Terms such as 'consumerism', 'corporatism', and 'economism' suggest that some actors in society interpret neoclassical economics as highly ideological. Valuation is dealt within monetary terms as part of the neoclassical efficiency concept, implying that there is a kind of 'monetary reductionism'. Neoclassical economists make reference to 'correct' prices as part of Cost-Benefit Analysis and claim to be able to point out the 'best' or 'optimal' alternative for society from a resource allocation point of view. Proponents of other ideologies are seldom as precise in their reasoning and conclusions. Institutional economics as presented here is for the same reasons science as well as ideology but tries to deal with this 'fact' rather than deny it.

Neoclassical economics is part of positivism as a theory of science and it is argued that objectivity and value neutrality is possible. The neoclassical project from the 1870s onwards has in fact been an attempt to make economics a 'pure' science. Economics is regarded as separate and separable from politics. According to this view, science and university education could not be part of the development problems faced. Researchers and teachers at universities study various phenomena in a value-neutral way. They are looking for 'truth' and nothing else. Nobody can blame them for the development problems. Rather, politicians and perhaps business actors are responsible for such situations. This view tends to dominate in spite of all that has happened in the theory of sciences in terms of a new interest in the subjective aspects of research and human behaviour. Hermeneutics, narrative analysis (Porter Abbott 2002), and social constructivism are examples of this.

Institutional economists, such as Gunnar Myrdal, have questioned the neoclassical position:

Valuations are always with us. Disinterested research there has never been and can never be. Prior to answers there must be questions. There can be no view except from a viewpoint. In the questions raised and the viewpoint chosen, valuations are implied.

·Our valuations determine our approaches to a problem, the definition of concepts, the choice of models, the selection of observations, the presentations of conclusions—in fact the whole pursuit of a study from beginning to end.

(Myrdal 1978, pp. 778–9)

Myrdal furthermore speaks of the necessity therefore 'in any scientific undertaking of stating clearly and explicitly, the value principles which are instrumental'. Consciousness about value issues—or in the present terminology—about how ideology might affect a study or lecture then becomes a quality criterion of good research and education.

The reasons to replace neoclassical monopoly with pluralism can now be better understood. If each paradigm or scientific perspective is coloured by values and ideology, then limiting research and education to one paradigm at a university department is not compatible with normal ideas about democracy. Science and universities should not take a stand for one particular ideology, such as the neoclassical 'market and economic growth' ideology, at the expense of all other ideological options. Science should rather illuminate an issue in relation to various possibly relevant ideological orientations. Ezra Mishan (1971) has argued that the use of cost–benefit analysis for decision-making in society should be conditioned upon a consensus among citizens about the rules of valuation built into this analysis. According to Mishan, the polarized debate over environmental issues implies that this consensus no longer exists (Mishan 1980). A Norwegian economist similarly identifies the ideology of cost-benefit analysis as being close to economic growth in GDP terms (Johansen 1977). Such an ideological commitment is acceptable for some but not for others.

The case of neoclassical economics suggests that the mentioned failures cannot be seen in isolation. Paradigm is combined with ideology and both are connected with institutions. Perhaps one should speak of a cluster with elements of paradigm, ideology, and institutions in competition with other clusters of paradigm-ideology-institutions. Neoclassical economics and neo-liberalism are compatible in many ways and with the present growth-oriented market economy as the main institutional arrangement. Proponents of this 'cluster' systematically avoid many issues that are seen here as fundamental. 'Market failure' and 'government failure' as recognized by neoclassical economists will here be regarded as only a subset of possible 'institutional failures'. As part of a broader view, the 'firm' or business corporation as institution, while celebrated as an 'engine of growth', may not in all respects be well adapted to present needs. The size of business corporations is also an issue, according to David Korten (2001) who questions the growth of transnational companies in power terms and thereby in relation to democracy.

Also organizations other than business companies, for instance civil society organizations, may fail. They may—as much as many other establishment actors—avoid an open discussion about paradigms in economics and ideology. Individuals may fail in professional and other roles and the total lifestyles of individuals can be problematic from an environmental point of view. Fortunately, there are also examples of 'success' and 'good practice' in the above respects.

Implicit in the previous arguments is that sustainable development can be seen as an ideological orientation in competition with other ideological orientations. If the conceptual framework of neoclassical economics is not helpful in guiding us towards sustainable development, then this becomes a reason to refer to 'paradigm failure'. On the basis of a specific definition of sustainable development, it may furthermore be argued that traditional ideologies such as liberalism and socialism have failed to the extent that they have not yet sufficiently 'internalized' values connected with sustainable development.

Sustainable Development

Any distinction between 'success' and 'failure' presupposes a specific value or ideological orientation. Although not completely clear, 'sustainable development', as defined by the Brundtland Commission (World Commission on Environment and Development 1987), and as further articulated during the Rio process and preparations for the Johannesburg conference, is such an ideological orientation. Implicit in it is a distinction between sustainable development and unsustainable development. A considerable part of activities in Sweden or any other country now follow an unsustainable development path and the challenge ahead is to bring an increasing share of activities closer to a sustainable development path or trajectory.

Building on the Brundtland Commission, it is here suggested that sustainable development is:

- Understood and measured in multidimensional terms where cultural and social dimensions in a broad sense, physical and ecological dimensions of various kinds, and financial or monetary dimensions of various kinds are considered;
- Built on ethical principles where not only present generations in the 'home country' but also present generations in other countries as well as future generations at home and globally are involved, and ethics in relation to non-human forms of life is also considered;
- Built on a precautionary principle, for instance in the sense that there is an ambition to avoid irreversible damage to people and ecosystems even in cases where such negative impacts are uncertain; and
- Built on normal ideas about democracy, such as participation and open access to information.

Non-degradation of the state of the environment or of the natural resource base is a primary condition for sustainable development in broader social and monetary dimensions to come true. As an example, focus can be on non-degradation of the ozone layer and other parts of the atmosphere and hydrosphere that are essential for life on earth. Non-degradation can similarly be an ambition for parts of the built environment and cultural heritage in various parts of the world. But also in relation to the preservation of, say, old buildings, there are difficult ethical and ideological issues involved. Building a new infrastructure such as a dam for electricity and other purposes or a highway for

	Present Generations	Future Generations
'Home region'	X	X
Other regions	X	X

Figure 4.1: A Narrow and a More Inclusive Idea of Ethical Considerations

transportation is normally at the expense of ecosystem services and other functional aspects.

When considering planning options for a specific region from an anthropocentric point of view, ethical imperatives should include relationships between the present and future generations in the same region, between the present generation in the home region and present generations outside the home region, and finally between the present generation in the home region and future generation outside the home region (see Figure 4.1). A focus on a region or river basin should be part of a broader perspective in time and space.[1]

The democracy aspect of sustainable development, too, needs to be elaborated upon. Agenda 21, as one of the agreements from the Rio conference, emphasizes local democracy and mobilization of civil society as leading principles. Democracy starts ideally from such a local level with as many persons as possible involved in an interactive learning process. All concerned should have access to essential information and, in addition to dialogue and co-operation, there should be a fair competition between various ideas about a desirable future in local and global terms. Democracy means pluralism and recognition that there are

---

[1] An early attempt to formulate ecological imperatives for public policy in these terms can be found in Söderbaum 1982. The precautionary principle is addressed in Harremoës et al. 2002.

different ideological orientations in society. Groups that at any time hold ideological orientations that depart from the dominant ideology should be encouraged as long as their opinions and behaviour do not negate democracy itself. Democracy furthermore includes a mutual control aspect. Citizens and specific professional categories such as journalists should watch all kinds of activities and point to behaviour that depart from laws and established norms in society. Specific civil society organizations (so-called non-governmental organizations [NGOs]) may take a leading role in attempts to speed up moves towards a sustainable development-path (Edwards et al. 2001).

For the moment, I will only make one observation. Sustainable development is essentially discussed as a vision for development at the societal or macro level. But in my understanding, to make sustainable development possible for society as a whole, individuals and organizations at the micro level too have to behave and act in a way that is compatible with sustainable development. This means, as an example, that business companies have to replace their one-dimensional focus on monetary profits, shareholder value, and bonus systems in monetary terms for Chief Executive Officers (CEOs) and board members with some multidimensional ideas of performance. Something has happened in positive terms, for instance the recent debate about social responsibility of business and the institution of environmental management systems. Some companies take a leading role in this new development but the main idea of business policy and practice is still one of more or less institutionalized 'monetary reductionism'.

If sustainable development is connected with moves towards a strengthened democracy as suggested above, then a lot remains to be done also at the level of organizations. As an example, small shareholders in big companies could hardly be happy with the present state of affairs in terms of possibilities for environmental or other policies of the company. And we are all stakeholders to the extent that the activities of a company influence the global environment.

Ecological Economics as a Conceptual Framework for Sustainable Development

Ecological economics can be defined as 'economics for sustainable development' or 'economics in the spirit of Agenda 21'. Ecological economics is built on a commitment to work for sustainable development as a basic value premise. It is based upon the assumption that no social science can claim value neutrality. While emphasizing economics and business management, the ecological economist is eager to learn about useful ideas for the purposes of sustainable development from any other discipline. Interdisciplinary approaches and pluralism, that is open-mindedness to alternative theoretical or methodological perspectives, are therefore further characteristics of ecological economics.

Assuming now that 'business as usual' in terms of paradigm, ideology, and institutional arrangements will not be enough to guide us towards sustainable development, then we need to consider ways of consciously modifying or changing our mental maps or conceptual frameworks. The idea is then one of presenting a conceptual framework that is understood by actors in various roles and positions as being more useful in furthering the idea of sustainable development. In what follows, I will point to concepts at the micro level that together form the embryo of a new microeconomics, which in turn opens the doors for new thinking at the macro level. In many ways these ideas are in line with the conceptual framework that is emerging as part of the previously indicated water policy discourse.

Political Economic Person and Political Economic Organization

Economic Man is the cornerstone of neoclassical economics. Economic Man is exclusively related to a market context and the individual is essentially seen as a consumer maximizing utility, subject to a monetary budget constraint. While it is difficult to question a statement that man maximizes utility in some sense, it is equally true that such a statement is rather empty and uninteresting in relation to present environmental and development issues. Our interest is rather to find out how individuals differ with respect to their ideological orientations and lifestyles. To what extent is the ideological orientation of a specific individual compatible with the sustainable development ideas as previously defined? As an alternative more in line with institutional theory, a Political Economic Person (PEP) is proposed,[2] i.e. an individual with many roles (professional, consumer, citizen, parent, etc.) and relationships who is guided by a political or ideological orientation. The individual has an identity and is positioned in and interacts with a context that is social, institutional, physical (man-made), and ecological.

While self-interest is a dominant feature of Economic Man assumptions, it is here assumed that a healthy individual has a strong ego but that he or she at the same time is able to more or less internalize the interests of others. Amitai Etzioni speaks of an 'I & We Paradigm', (Etzioni 1988) according to which each individual is part of a number of 'we-categories'. As a person, I may be concerned about my family as one we-category, colleagues at my work-place as another we-category, and my home town, my region, and to some extent even the global society as other we-categories. Our Political Economic Person is a responsible actor who through networks and organizations can influence development processes at various levels.

In business management literature, alternatives to the neoclassical profit-maximizing firm have been available for some time, e.g. a 'stakeholder model' of organizations. Stakeholders are those concerned or those who have something 'at stake' in relation to a specific decision situation or the activities of an organization. Pointing to different categories of stakeholders such as shareholders, customers, employees, board members, CEOs, people living in

---

[2] Political Economic Person and other parts of this alternative microeconomics is outlined in Söderbaum 1999 and 2000 and is discussed for instance by Jakubowski (1999, 2000). A concept that is close to PEP, Homo Politicus has been suggested by Faber et al. (2002).

the neighbourhood, etc., is already a step forward in admitting that some conflicts of interest are normally involved in decision-making about investment projects or operational activities. As part of presentations of the stakeholder model, there may still be a tendency to pack together all shareholders, all employees, etc., into homogenous categories even though we all know that not all shareholders (or all employees) have the same interests.

As a complementary model and a way of allowing for such differences between individuals as actors, the Political Economic Organization (PEO) is proposed. A PEO is composed of a number of individuals as PEPs, which means that the organization is regarded as 'polycentric', each individual being represented with her or his particular roles, relationships, and ideological orientation. In a business company or other organization that takes steps in a Green direction by becoming certified according to ISO 14001 and in other ways, some individuals take the lead and become 'environmental entrepreneurs' while others are followers.

While a 'firm', according to neoclassical theory, is presented as something separate and separable from other firms and from consumers, the present network approach suggests that the same individual is an actor within or in relation to more than one organization or network and that individuals as well as organizations may cooperate for specific purposes (e.g. the mentioned 'I & We Paradigm'). The actor (individual or organization) is embedded in various cooperative relationships and networks with other actors. Two organizations as actors may compete in relation to some activities (for example, technological development and market penetration) and cooperate in relation to others (for example lobbying activities in relation to regulatory entities, such as the European Union).

Concepts of Economics and Efficiency

Sustainable development as previously described is not a completely clear vision but it suggests certain directions for development at the macro and micro levels. Ideas connected with sustainable development may more or less influence the ideological orientations of specific actors in government, business, and civil society. To the extent that sustainable development with its imperatives of democracy is taken seriously, a multidimensional and ideologically open idea of economics and efficiency will emerge. Monetary dimensions will still be important in our market economies but attempts to reduce non-monetary impacts to their alleged monetary equivalents will lose in terms of relevance and legitimacy. In addition to the distinction between monetary and non-monetary impacts, a distinction is made between variables expressed as flows (referring to periods of time) and as positions or states (referring to points in time). This is illustrated in Figure 4.2.

Among examples of monetary flows (category I in Figure 4.2), GDP, the turnover and profits of a business company, and the salary of an employee can be mentioned. The assets and liabilities in monetary terms of a business company at the beginning or end of an accounting period exemplify monetary positions (category II). The discharge of a pollutant such as mercury to a nearby lake is an example of a non-monetary flow (category III) while the content of mercury in fish (as measured in ppm, parts per million) caught in the lake at a specific place and point in time is a non-monetary position (category IV). All four categories of impacts should be kept separate in economic analysis. In relation to sustainable development, non-monetary variables play a crucial role and especially parameters in terms of positions or

	Flow (Referring to a Period of Time)	Position (Referring to a Point in Time)
Monetary	I	II
Non-monetary	III	IV

Figure 4.2: Classification of Variables in Measuring Resources and Impacts for Purposes of Economic Analysis

states are essential for judgements about changes in welfare. How is the stock of fish of various species changing in a lake from one point in time to another? What happens to groundwater quality over time at a place? Such series of positions referring to relevant objects of description will tell us a lot about changes in the state of the environment, and changes in the health of human beings or of ecosystems. In this part our argument is in line with the focus on the 'state of the world' in the annual reports by the Worldwatch Institute (see, for example, Brown 2001).

Cost-Benefit Analysis with its one-dimensional and ideologically closed ideas of values and efficiency at the societal level will no longer be accepted. 'Value' is regarded as equal to 'monetary value' and the analysis is essentially one in terms of monetary flows (category I). From the point of view of democracy, the scepticism in relation to Cost-Benefit Analysis expressed by the World Commission on Dams (WCD 2000) represents an important step forward. In a democracy, scientists (economists) have no right to dictate correct values for purposes of societal resource allocation (Söderbaum 2001a). As an alternative to this 'monetary reductionism', the purpose should instead be one of illuminating an issue for actors of different ideological orientations. Here approaches such as Positional Analysis in terms of multidimensional impact profiles for alternatives considered and a 'matching' idea of decision-making is an option (Söderbaum 2000). The analyst who takes democracy seriously has to consider more than one ideological orientation and formulate his conclusions accordingly.

The ambition of neoclassical economists is to include as many impacts as possible in their monetary analysis. David Pearce, for instance, refers to 'Total Economic Value', which includes 'Actual Use Value', 'Option Value', and 'Existence Value' (Pearce et al. 1989, p. 62). Others prefer the term 'capital' and speak of 'total capital' including 'man-made capital', 'social capital', and 'natural capital'. Some degradation of natural capital can then be made legitimate by increases in other kinds of capital. According to this philosophy everything can be 'traded' against everything else.

The neoclassical conjurer is able to reduce multidimensional complexity to one-dimensional simplicity. Assuming that this was possible there is still the issue of choosing the prices at which trade takes place. Again, there is no acceptable solution for those of us who take democracy seriously.

Views of Market and Non-market Relationships

In any attempt to outline a microeconomics more in line with Sustainable Development, our ideas about markets also have to be involved (Figure 4.3). Here neoclassical economists stick to mechanistic ideas about supply and demand for specific commodities. This model focuses on prices in monetary terms and quantities exchanged and it certainly has some explanatory value. But since the efficiency of markets in allocating resources is an open issue as previously discussed, the tendency to extend monetary business and market thinking to new areas has to be scrutinized carefully. Will we get a better world by, for instance, exchanging 'water use rights' (e.g. Simpson and Ringskog 1997) or 'pollution rights' in markets?

Second, for that part of activities where markets can play a positive role, other models of market exchange than the supply–demand model have to be developed and considered. A market can be regarded as a multifaceted relationship between market

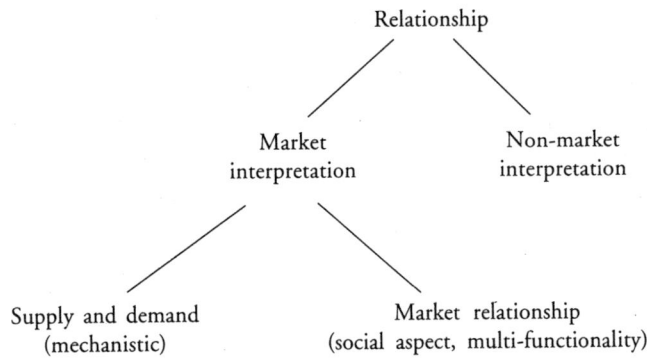

Figure 4.3: Relationships in Market or Non-market Terms

actors where social aspects and history plays a role. As an example, business-to-business market relationships are often better understood if one focuses on market actors (as Political Economic Persons or Political Economic Organizations) and how they relate to each other (commitments, trust, etc.) rather than exclusively on commodities sold or bought at specific prices (Ford 1990). As part of such social relations, the power positions and ideological or ethical orientation of each market actor become relevant. Prices are judged in relation to the ideological orientation of each market actor in terms of their 'fairness' and 'reasonableness', for instance, and not just as a matter of maximization of self-interest by atomistic actors.

The recent debate about the World Trade Organization (WTO) furthermore suggests that there is a 'one-commodity fallacy' in neoclassical microeconomics as well as neoclassical international trade theory. Trade in a national or international context is about much more than quality and quantity of product with connected prices. In the case of agriculture, it has been recognized as part of WTO negotiations that the welfare and culture of individuals and collectivities are influenced in a number of ways, local landscapes and other aspects of natural resources included. 'Multi-functionality' has emerged as a key concept in understanding this complexity (OECD 2000). Others point to a 'public goods' aspect of almost all activities and exchange relationships (Kaul et al. 1999). In my judgement, traditional international economics textbooks will have to be rewritten to become relevant to the world in which we live. But I also recognize the ideological character of present texts and therefore the reluctance in some circles to open the doors for new thinking.

Actors, Agendas, and Arenas for Social and Institutional Change Processes

Institutional theory is helpful in understanding social change processes. An actor-agenda-arena approach based on PEP-assumptions points to the importance of ideological orientation and the theoretical perspectives or conceptual frameworks used in interpreting various phenomena. Any attempt to get closer to a sustainable development path will involve changes in conceptual framework, language, and connected interpretations. This can be illustrated by reference to the example of an increased number of business companies being certified according to ISO 14001. A business company is itself an 'institution' and for a long time has been interpreted by many as a 'profit-maximizing organization'. Neoclassical economics advocates such an interpretation, and business management literature and journalism tend to point in the same direction. At some stage, as a result of the initiatives of some actors connected with business, environmental management systems (EMS) such as ISO 14001 did appear on the scene as an 'institution' in itself. An increasing number of individuals understand the meaning of an EMS, implying that the institution is strengthened. But the fact that some companies are certified according to ISO 14001 may in turn change our understanding of the business companies being certified. A 'certified business company' is interpreted as being different from companies that are not certified. The institution of 'business company' is then understood in broader terms. It is not only a matter of monetary profits but also of environmental performance. One may speak of a competition between the 'old' interpretation and a newer one.

Three aspects of such institutional change processes are relevant, namely interpretation, legitimacy, and manifestation. An 'institution' becomes strengthened or more established to the extent that it becomes manifested in symbols or concrete behaviour among an increasing number of actors. 'Institutionalization' or 'deinstitutionalization' may take place through all these three processes. Institutionalization here refers to an institution that becomes strengthened and more established among actors, while deinstitutionalization refers to a situation where an institution over time is losing its support and finally may become out-competed by other institutions.

As another example, a specific version of ecological economics can become more institutionalized over time through manifestations in terms of international and regional organizations, journals, articles, conferences, educational programmes, professorships, and so on. Ecological economists contribute to public debate on environmental and development issues and each such contribution is part of a broader evolutionary process where the policies and actions of various actors are shaped.

The present actor–agenda–arena approach[3] can be characterized as having the following salient points:

- Political Economic Person assumptions
- Emphasis on relationships between actors who, with their specific ideological orientation or Agenda, appear on specific Arenas
- Emphasis on the conceptual and interpretative aspect of 'ideological orientation'
- Dialogue, search for consensus, conflict resolution, and other aspects of interactive learning
- 'Institution' and 'institutional change' are defined in interpretative, legitimacy, and manifestation terms
- An assumption of heterogeneity of 'ideological orientation' in each conventionally defined actor category (farmers, business leaders, university scholars, etc.)
- An assumption that actors search for commonality in ideological terms by building networks and alliances within and across conventionally defined actor categories.

Only the heterogeneity assumption will be further explained here. In neoclassical theory, more precisely public choice theory, an assumption of homogeneity is made concerning farmers as a category, bureaucrats as a category, etc. This positivistic theory is of some interest but our more normative and interpretative approach suggests that differences within each category (farmers, bureaucrats, or business-leaders) too are relevant. Some farmers

---

[3] See also Söderbaum 2001b.

(business leaders) are concerned about environmental issues while others are not. Actors with a similar ideological orientation but belonging—in conventional terms—to different actor categories may work together as part of a common sustainable development strategy.

'Ecological Modernization' is not Enough

All kinds of measures to improve environmental and social performance at different levels should be encouraged and small changes in the direction of a sustainable society could be part of more radical transformation processes. But the obstacles in front of us should not be underestimated. Here I will make an admittedly simplified distinction between three ideological orientations—with connected response patterns—in relation to environment and development:

- 'Business as usual': According to this view, statements about the existence of environmental and social problems are generally exaggerated. To the extent that such problems exist, they can easily be handled within the scope of a continued emphasis on economic growth, technological innovations, and global market penetration. No change in dominant paradigm, ideology, or institutional framework is needed.
  Response pattern: 'If we do not speak about the problems, then perhaps they do not exist.' 'To the extent that they exist, they will be taken care of by the in-built mechanisms of our present market economy.' 'Public Relations campaigns and lobbying will make people focus on traditional parameters and forget about environmental problems.'
- 'Ecological modernization' (e.g. Hajer 1995): Humanity faces environmental and social problems of a serious and, in some respects, new kind. In all organizations, management systems have to be modified to allow for this new situation. While action is needed at all levels, from the individual,

organization, local government, and national government to the global level, presently dominating paradigms, ideologies, and institutions need only be modified and 'modernized' to allow for the new situation.

Response pattern: 'Yes, there are problems but don't worry, things are under control.' 'The social responsibility of business will be reconsidered.' 'Voluntary agreements, environmental management systems, environmental labelling, etc. will do it.'

• 'Major shifts in paradigm, ideology, and institutional framework': Modifying paradigm, ideology, and institutional framework may get us closer to a sustainable development path but this will not be enough. Sustainable development as an ideological orientation has to be taken seriously and a 'major shift in thinking' is indispensable.

Response pattern: 'We need other conceptual frameworks in economics as part of a pluralistic and democratic philosophy.' 'Competition is preferred to the global neoclassical monopoly or cartel at university Departments of Economics.' 'Individuals and organizations alike need to reconsider their ideas of progress as actors privately, professionally, and in society.'

An example of the 'business as usual' attitude is an advertising campaign by the Confederation of Swedish Enterprise (May 2002), in which welfare is exclusively connected with economic growth in GDP terms and where the main concern is Sweden's position when compared with other countries in a 'welfare-league' as measured in GDP terms. Not one word is said about the environment or broader ideas about welfare such as sustainable development. This is not to say that the Confederation of Swedish Enterprise completely neglects the debate about sustainable development—they have in fact recently employed a person to take care of these issues—but to judge from the confederation's advertising, web site etc., sustainable development is not a big issue for them.

The ideological orientation of 'ecological modernization' is described as follows:

Ecological modernization . . . uses the language of business and conceptualises environmental pollution as a matter of inefficiency, while operating within the boundaries of cost-effectiveness and administrative efficiency. (Hajer 1995, p. 31)

Ecological modernization explicitly avoids addressing basic social contradictions. [It] does not call for any structural change but is, in this respect, basically a modernist and technocratic approach to the environment that suggests that there is a techno-institutional fix for the present problems. (Ibid., p. 32)

In the most general terms, ecological modernization can be defined as the discourse that recognizes the structural character of the environmental problematique but none the less assumes that existing political, economic, and social institutions can internalise the care for the environment. (Ibid., p. 25)

This ideological orientation is exemplified by the World Business Council for Sustainable Development (WBCSD), which is a coalition of 140 international companies working for economic growth and sustainable development. While 'sustainable growth' in GDP terms seems to be a top priority, there are also some openings in publications from this organization:

A growing number of business leaders realize that to achieve market success they must honor a changing array of environmental and social responsibilities . . . As business leaders, we understand and respect the workings of the market. But we know that the market is not some ruling entity separate from human activities. (WBCSD 1997, p. 56)

The media and consumers are becoming too sophisticated to allow companies to pretend; they expect real corporate action. (Ibid., p. 51)

Proponents of our third ideological orientation, 'major shifts in paradigm, ideology, and institutional framework', can be found in many professional and other categories but are perhaps best associated with civil society organizations, for instance the French Attac movement with actors such as Susanne George (2000) and René Passet (2000). George and Passet both point to neo-liberalism and neoclassical economics as part of the problem, as does David Korten in his book *When Corporations Rule the World* (2001). In relation to water issues, feministic perspectives may add to our understanding (Shiva 2002; Tortajada 2000). Fiction writers can provide a different conceptual and ideological perspective (Roy 2001) and thereby contribute to our understanding of issues such as the impacts of dam building. As exemplified by Jeremy Legget's study of climate change negotiations (1999), persons who combine scientific knowledge with journalism can similarly make the behaviour of various actors on the local or international scenes more visible.

While there is still some room for the third more radical interpretation of sustainable development, actors with 'business as usual' or 'ecological modernization' attitudes tend to dominate the scene these days. Transnational companies and politicians with a neo-liberal orientation have been successful in defining the problems and influencing the development dialogue. There is even a tendency towards the replacement of traditional ideas of business being regulated by national governments and through international agreements between national governments by a situation where business is controlling and regulating national governments. The Transatlantic Business Dialogue (TABD) is an example of this trend. Transnational corporations in the US and in the European Union claim specific rights to accept or not accept political proposals that concern them and also have specific channels to the US government and administration and the Commission of the European Union, respectively.

More power to business organizations that are still essentially governed by monetary principles will neither give us a sustainable society at the regional nor at the global level, nor is it in the interest of business itself. A global world ruled by the international business community is for many of us not much better than the Soviet-type planned economies of the past. Democracy is based on some division of power and does not permit any coalition of organizations to take over leadership and control. Instances of failure and mismanagement by 'big business' in the recent past have led to reactions by civil society in different parts of the world. It is quite probable that at some stage more power to business will undermine the power of business itself.

It is not my intention here to say that business is the only culprit or to make general statements about business actors. Rather, I would like to return to the previous actor-agenda-arena framework and its heterogeneity assumption. Proponents of a radical interpretation of sustainable development can be found in any category of actors and hopefully also among professionals in business. Similarly, universities and professionals as actors within universities differ in their ideological orientations. And a lot remains to be done at universities before one can claim that environmental and development issues are taken seriously. The 'business as usual' attitude is as common in university circles as elsewhere.

Integration of Policy Areas

One theme in this essay has been the need for broader approaches to development policy. In the case of water, the 'international water community' certainly has taken important steps in these directions. It is also recognized that problems related to water quality and availability are linked with all kinds of activities and sectors in society. Water policy has to be 'integrated' with agricultural policy, industrial policy, etc., to reduce pollution from various activities and thereby improve water quality. In this sense, to solve water problems, there is a need for a 'non-water policy' (Figure 4.4). A large number of chemicals such as

	Water	Non-water
'Home region'	X	X
Other regions	X	X

Figure 4.4: A Narrow and a More Inclusive View of Water Policy

synthetic hormones, antibiotics, and pesticides (e.g. Wolfe 2002) continue to pollute surface water and groundwater in different parts of the world in the name of rationality and efficiency. Could these problems be handled at all within the scope of a traditional conceptual framework and development ideology? An increasing number of actors furthermore understand that there is a global aspect of local activities. But at the same time, a lot remains to be done in practical terms. And practice should not be limited to sewage treatment but should include far-reaching preventive measures.

This means that actors at local and national levels need to engage in formulating their 'global water and non-water policies' (Figure 4.4), which in turn points to the need for cooperation in different forms and for international agreements (Porter et al. 2000). While there is always a global aspect of water policy, it is equally true that water policy and water management begin at the local or regional level. A 'regional' level may refer to administrative borders or to a river basin or catchment area. For river basins, sustainable development can be a guiding principle as exemplified by a study for the Ganges–Brahmaputra–Meghna Region (Ahmad et al. 2001).

Recommendations for Johannesburg and Beyond

Strengthening democracy. Social change processes reflect an interplay and, at the same time, a power game between individuals and various collective entities. Listening to many voices and learning from many sources will not only strengthen democracy but also in most cases lead to better problem-solving. Problems related to environment and development are extremely complex and many viewpoints and perspectives should therefore be considered. In the case of water policy and management, voices from civil society, could be important.

University research and education should be seen as part of politics and democracy. Social science—and economics in particular—

cannot be separated from values and ideology: we are all—scholars and other actors—Political Economic Persons in the sense indicated above. Among the criteria for being 'scientific' is the criterion that one openly discusses how values are involved in a study. According to Amartya Sen (1987), economics has for some time followed an 'engineering tradition' rather than its 'ethical tradition' and it is now time to reverse this trend. The economic growth debate over the years (Friman 2002) or the debate about international trade theory and the WTO is not just a matter of truth in some scientific sense but as much a matter of ideology. Such debates cannot be left to economists whose main interest too often seems to be to protect their neoclassical paradigm. Other professionals and politicians have to participate in this dialogue rather than act as if the expertise of university scholars also includes ideology. It is a mistake to believe that science can easily be separated from politics. The concept of research policy should include a possibility for intervention against ideological and scientific monopolies at universities.

In the case of social sciences, the concept of paradigm-shift should be replaced by paradigm-coexistence. Since each paradigm is coloured by values and ideology, only pluralism can be accepted in a democratic society. This means that different theoretical perspectives connected with different ideological orientations will exist side by side. 'Paradigm-coexistence' rather than 'paradigm-shift' in the Kuhnian sense (Kuhn 1970) will be the natural condition. There may still be competition between advocates of various paradigms and as a result of this, 'change in dominant paradigm'.

It may be added here that at least two different perspectives are represented in the International Society for Ecological Economics, with its journal *Ecological Economics*. One is more a part of an 'interface' and 'engineering' tradition with Robert Costanza, Charles Perrings, and Carl Folke as representatives while another is in favour of a more radical interpretation of sustainable development. The Costanza group leaves neoclassical economics and ecology essentially intact and focuses on modifications to allow for concepts such as ecosystem services and ecosystem resilience (Folke et al. 2002). Ecosystem services are still measured in monetary terms as in neoclassical economics. Richard Norgaard (1989) and the present author can be mentioned as representatives of the other school that is more critical of neoclassical economics and emphasizes a need for pluralism. An interesting observation is that the European Association for Ecological Economics appears to be more radical than its US counterpart.

Further research is recommended in the area of actors, agendas, and arenas. In the introduction, reference was made to the Stockholm + 30 conference. While some of the Swedish contributions to the Johannesburg Summit emphasized efforts to understand 'decoupling' (of environmental problems from economic growth) (Azar et al. 2002) or resilience as mentioned above, i.e. rather impersonal mechanisms, one speaker, Mark Nerfin, as part of a panel discussion focused on actors and the possibility of self-assessment. In addition to other efforts, it is a good idea, he stated, to focus on your own person and organization by reasoning in terms of accountability and similar concepts. Is your lifestyle and the activities of your organization compatible with sustainable development? It is easy to point to the positive part—my university, for instance, being the first in Europe and perhaps internationally to be certified according to ISO 14001, or the undergraduate ecological economics programme that I am in charge of. However, the overall situation of the Mälardalen University still leaves a lot to be desired in this respect, as in other programmes we are still educating economists as if environmental and development problems do not exist or play a minor role.

Self-assessment is of course only part of the story. As students we can focus on other actors, their ideological orientation, roles and relationships, as well as institutional arrangements. Tape-recorded interviews with different actors related to an issue can facilitate self-reflection and sometimes dialogue and action. Here a 'responsibility trap' may be revealed in the sense that each actor tends to limit his or her responsibility as part of traditional ideas about specialization. 'Environmental issues are outside my concerns and belong to the table of Mr X.' In this way one gets an indicator of how far the policy of 'integrating' various areas has reached in practice.

Another actor category to be approached is that of politicians. What efforts have been made to internalize sustainable development into traditional political ideologies such as liberalism or social democracy? Is 'business as usual' or 'ecological modernization' the best way to describe such attempts? Is there a willingness to reconsider some of the more fundamental premises of present policies?

Like other actors, we are, as scientists and teachers, part of a democratic society and have an important critical role in relation to sustainable development policies. Such a critical role may be more easily accepted if one recognizes that there is no apolitical science. Approaches to management and decision-making are important parts of possible contributions from research as is impact assessment. And in our respective roles, attempts to influence the political agenda by raising the more fundamental issues of paradigm, ideology, and institutions at various arenas are crucial. In fact, the present international consensus on sustainable development among a large number of actors leaves a lot of room for efforts of this kind.

References

Ahmad, Q. K., A. K. Biswas, R. Rangachari, and M. M. Sainju (eds), 2001, *Ganges–Brahmaputra–Meghna Region: A Framework for Sustainable Development*, The University Press Ltd, Dhaka.

Azar, C., J. Holmberg, and S. Karlsson, 'Decoupling: Past Trends and Prospects for the Future', Swedish Environmental Advisory Council Report 2002: 2, Stockholm.

Brown, L. R., 2001, *State of the World 2001*, A Worldwatch Institute Report on Progress Toward a Sustainable Society, Earthscan, London.

Commission of the European Communities, 2001, *Environment 2010: Our Future, Our Choice—The Sixth Environment Action Programme*, Brussels.

Department for International Development (DFID), United Kingdom, Directorate General for Development, European Commission (EC), United Nations Development Programme (UNDP), The World Bank, 2002, *Linking Poverty Reduction and Environmental Management: Policy Challenges and Opportunities* (A contribution to the World Summit on Sustainable Development Process), DFID, London.

Edwards, M. and J. Gaventa (eds), 2001, *Global Citizen Action*, Earthscan, London.

Etzioni, A., 1988, *The Moral Dimension: Towards a New Economics*, Free Press, New York.

Faber, M., T. Petersen, and J. Schiller, 2002, 'Homo Oeconomicus and Homo Politicus in Ecological Economics', *Ecological Economics*, vol. 40, no. 3, pp. 323–33.

Falkenmark, M., 2000, *No Freshwater Security Without Major Shift in Thinking: Ten-Year Message from the Stockholm Water Symposia*, Stockholm International Water Institute, SIWI, Stockholm.

Folke, C. et al., Resilience and Sustainable Development: 'Building Adaptive Capacity in a World of Transformations', Swedish Environmental Advisory Council Report 2002: 1, Stockholm.

Ford, D. (ed), 1990, *Understanding Business Markets: Interaction, Relationships, Networks*, Academic Press, London.

Friman, E., 2002, 'No Limits: The 20th Century Discourse of Economic Growth', Ph. D. Thesis, Department of Historical Studies, Umeå University, Umeå.

George, S., 2000, 'A Short History of Neoliberalism: Twenty Years of Elite Economics and Emerging Opportunities for Structural Change', in B. B. Walden, N. Bullard, and K. Malhotra (eds), *Global Finance: New Thinking on Regulating Speculative Capital Markets*, Zed Books, London, pp. 27–35, Chapter 2.

Hajer, M. A., 1995, *The Politics of Environmental Discourse: Ecological Modernization and the Policy Process*, Clarendon Press, Oxford.

Harremoës, P., D. Gee, M. MacGarvin, A. Stirling, J. Keys, B. Wynne, and S. G. Vaz (eds), 2002, *The Precautionary Principle in the 20th Century: Late Lessons from Early Warnings*, Earthscan, London.

International Conference on Freshwater, Bonn 3–7 December 2001, Ministerial Declaration, in http://www.water-2001.de/

Jakubowski, P., 1999, *Demokratische Umweltpolitik*. Eine institutionenökonomische Analyse umweltpolitischer Zielfindung, Peter Lang Verlag, Frankfurt am Main.

————, 2000, Political Economic Person contra Homo Oeconomicus, Mit PEP zu mehr Nachhaltigkeit, List Forum für Wirtschafts, und Finanzpolitik, Band 26, Heft 4, pp. 299—310.

Kaul, I., I. Grunberg, and M. A. Stern (eds), 1999, *Global Public Goods: International Cooperation in the 21st Century*, United Nations Development Programme (UNDP), Cambridge University Press, Cambridge.

Korten, D. C., 2001, *When Corporations Rule the World* (second edition), Kumarian Press, West Hartford.

Kuhn, T. S., 1970, *The Structure of Scientific Revolutions*, University of Chicago Press, Chicago.

Johansen, L., 1977, 'Samfunnsökonomisk lönnsomhet: En dröfting av begrepets bakgrunn og innhold', Industriökonomisk Institut, rapport Nr 1, Tanum-Norli, Oslo.

Legget, J., 1999, *The Carbon War: Global Warming and the End of the Oil Era*, Penguin, London.

Ministry of the Environment, 2002, *Stockholm Thirty Years On: Progress Achieved and Challenges Ahead in International Environmental Co-operation*, Stockholm.

Mishan, E. J., 1971, *Cost–Benefit Analysis*, Allen & Unwin, London.

————, 1980, 'How Valid are Economic Evaluations of Allocative Changes?', *Journal of Economic Issues*, vol. 14, no. 1, March, pp. 143–61.

Myrdal, G., 1978, 'Institutional Economics', *Journal of Economic Issues*, vol. 12, no. 4, December, pp. 771–83.

Norgaard, R. B., 1989, 'The Case for Methodological Pluralism', *Ecological Economics*, vol. 1, no.1, pp. 37–57.

OECD (Organization for Economic Co-operation and Development) Directorate for Food, Agriculture and Fisheries, Trade Directorate, 2000, *Multifunctionality: Towards an Analytic Framework*, OECD, Paris.

Passet, R., 2000, *L'illusion néo-liberale*, Fayard, Paris.

Pearce, D. W., Aril M. Barbier and E. B. Barbier, 1989, *Blueprint for a Green Economy*, Earthscan, London.

Porter, A.H., 2002, *The Cambridge Introduction to Narrative*, Cambridge University Press, Cambridge.

Porter, G., J. W. Brown, and P. S. Chasek, 2000, *Global Environmental Politics* (*third edition*), Westview Press, Boulder.

Roy, A., 2001, *Priset för att leva* (*The price of survival*), Nya Doxa, Nora.

Sen, A., 1987, *On Ethics and Economics*, Basil Blackwell, Oxford.

————, 1999, *Development as Freedom*, Alfred A. Knopf, New York.

Shiva, V., 2002, *Water Wars, Privatization, Pollution and Profits*, Pluto Press, London.

Simpson, L. and K. Ringskog, 1997, *Water Markets in the Americas* (*Directions in Development*), World Bank, Washington, D.C.

Söderbaum, P., 1982, 'Ecological Imperatives for Public Policy, Ceres', *FAO Review for Agriculture and Development*, vol. 15, no. 2, pp. 28–30.

———, 1999, 'Values, Ideology and Politics in Ecological Economics', *Ecological Economics*, vol. 28, no. 2, pp. 161–70.

———, 2000, *Ecological Economics: A Political Economics Approach to Environment and Development*, Earthscan, London.

———, 2001a, 'Neoclassical Economics, Institutional Theory and Democracy: CBA and Its Alternatives', *Economic and Political Weekly*, vol. 36, no. 21, pp. 1846–54.

———, 2001b, 'Business Corporations, Markets and the Globalisation of Environmental Problems', in Virpi, H., M. Forsgren, and H. Håkansson (eds), *Critical Perspectives on Internationalisation*, Elsevier Science/Pergamon, Oxford, pp. 179–200.

Tortajada, C. (ed), 2000, *Women and Water Management: The Latin American Experience*, Oxford University Press, New Delhi.

Tortajada, C., A. Shady, and I. Al-Baz (eds), 2002, 'Dams, Energy and Regional Development', *International Journal of Water Resources Development* (*Special Issue*), vol. 18, no. 1, March.

UNEP, 2002, *Global Environment Outlook 3*, Earthscan, London.

Water Science and Technology, 2002, *Water Security for the 21st Century—Building Bridges through Dialogue* (Proceedings of the 11th Stockholm Water Symposium), vol. 45, no. 8.

Wolfe, P., 2002, Viewpoint, '*Water & Wastewater International*', April.

World Business Council for Sustainable Development (WBCSD), 1997, *Signals of Change: Business Progress Towards Sustainable Development*, Geneva.

World Commission on Dams, 2000, Dams and Development. A New Framework for Design-Making, Earthscan, London.

World Commission on Environment and Development, 1987, *Our Common Future*, Oxford University Press, Oxford.

5

Rat Catching in Sustainable Development

Alexander Gillespie

Introduction

Sustainable development is a term with many connotations and a subject that has been written about extensively. My broad contention is that there are two parts to the concept of sustainable development: an ethical component[1] and a political[2] component. These two components make up the core of the debate about what sustainable development should mean. These components are not related to the specifics of what is, or is not, actually sustainable. That debate—ultimately one of environmental limits—is to be found in the specific issues at hand. For example, the environmental limit for sustainability is to be found in the chlorine loading in the ozone layer (the point at which the impact outweighs the absorption capacity), or the carbon loading in the atmosphere (the point at which the increase outweighs the ability of ecosystems to respond naturally), or the point when more fish are taken from a stock and the replenishment level begins to dramatically fall, etc. As such, the real bottom line of what is or is not sustainable is located in each individual arena.

[1] See Gillespie, A., 1997, *International Environmental Law, Policy and Ethics*, Oxford University Press, Oxford.
[2] See Gillespie, A., 2001, *The Illusion of Progress: Unsustainable Development in International Law and Policy*, Earthscan, London.

In terms of the overall philosophical debate about sustainable development, a different set of issues arises, which are not about specifics, but about deeper and wider background concerns. Against this background, it is important to realize that sustainable development, as an idea, is about the environmental, social, and economic bases of development, where the ideal is to synthesize each one of these into an overall package.

It has been contended that sustainable development is not a new idea, and that ultimately it adds little to the current problems facing humanity. In terms of the politics and ethics of sustainable development, I do not disagree. However, in terms of this debate, about the overall merits of the idea of sustainable development, the discounting view is mistaken and fails to understand the context from which the problem originates. This context shows how development with only an economic focus, firstly, failed the social side of the equation. Thereafter, development with only an economic and social focus failed the environmental side. As such, only now do we recognize that sustainable development, as a yardstick (it can never be anything more) is about three things: the environmental, social, and economic foundations of development. Each one has an independent standing and cannot be subsumed beneath the auspice of either of the other two. It was precisely because of attempts to do so that the earlier failures occurred.

At this point, an attempt will be made to address some important questions raised by others, which implicitly question some of the underlying themes of sustainable development. This questioning has been best done by Bjorn Lomborg in his *The Sceptical Environmentalist: Measuring the Real State of the World*.[3] After consideration of possible topics relating to sustainable development, addressing some of the concerns raised in this book would be useful. This is necessary due to the large amount of publicity and support this text has attracted. In addition, the

[3] Lomborg, B., 2001, *The Sceptical Environmentalist: Measuring the Real State of the World,* Cambridge University Press, Cambridge.

willingness, with which it has been swallowed by many, is not surprising given the outward appearance of well researched, comprehensive, and value-free text. The text is not all of these things. Rather, despite clearly good intentions and containing many conclusions, it is a text that implicitly threatens many of the key objectives of any kind of meaningful sustainable development. As such, it needs to be answered.

Scepticism

Who are the Sceptics?

Before examining the merits of the *Sceptical Environmentalist*, it is necessary to say a little about the application of philosophical terminology and tools in Lomborg's text. Philosophical scepticism questions our cognitive achievements, challenging our ability to obtain reliable knowledge. This is a very useful and worthwhile goal, for if we are to seek the truth in a responsible manner, we need to confront all challenges and difficulties in which no defensible answer is available. Scepticism can be used for good or bad purposes. It can be used to challenge or support the status quo, or future ideals of where society should be trying to reach. However, if recklessly utilized, it may cause more damage to an area than benefits. Lomborg utilizes his scepticism to defend the status quo, and challenge alternative visions of a future society presented by those who he attacks. In addition, due to the way that this debate has unfolded, the final result will probably reflect an overall damage to the debate about sustainable development, rather than getting any closer to the 'truth' on such matters.

Scepticism, as a philosophical school, emerged in Greek philosophy, and arguably, its earliest (and best) advocate was Socrates, who possessed the ability to question and undermine any dogmatic assertion that was put to him, insisting that wisdom consisted in awareness of the extent of one's own ignorance. The targets of Socrates were those who held dogmatic views, about things they really knew nothing about.[4] For this, Socrates was known as the rat-catcher of Athens, as he would attempt to expose those who professed to know more than they actually did.

As such, Socrates' wisdom was to show that it was to possess greater knowledge by acknowledging our uncertainties about what we do not know and acting accordingly, rather than argue mistakingly from what we do not know. This realisation has direct applicability for the *Sceptical Environmentalist*.

Scepticism and Ideals for the Future

Socrates' contribution to this debate did not end with his pointing out uncertainties of others, and leaving the matter there. Rather, he emphasized the importance of knowledge that was properly grounded or tethered, and explained how such knowledge was possible, suggesting that it was necessary for seeking philosophical excellence. As such, it was ultimately possible to gain some type of truth, and (for this instance) build society towards it. In many ways, this aspect becomes a debate about visions of what society should be trying to achieve in the future. A good example of this is provided by some of the more memorable debates between the Sceptics and the Stoics in antiquity. In a very rough explanation, the Stoics built many (but not all) of the foundations which ultimately evolved into what we now recognize as human rights. Their primary adversaries were the Sceptics, who argued against ideas such as 'universal' ideals in which all people were equal under one set of laws, due to problems of relativism.[5] Of course, this debate continues 2000 years on in the human rights literature (as it has done through Western philosophy in general since the Renaissance) with the

---

[4] See Annas, J. and Narnes, J. (eds), 1985, *The Modes of Scepticism: Ancient Texts and Modern Interpretations*, Cambridge University Press, Cambridge; Barnes, J., 1985, *The Toils of Scepticism*, Cambridge University Press, Cambridge.

[5] Gillespie, A., 2001, 'The Roots of the Human Rights Debate in Antiquity', *Netherlands Journal of Human Rights*, (1999) vol. 17: 3, pp. 233–58.

language of cultural relativism and all of its implications bandied about. The intention here is not to discuss the merits of the human rights debate, but only to point out that much of the sceptical tradition is to focus negatively on ideal models for society. Such discussions have large implications for the debate about sustainable development in general, as much of the vision of meaningful forms of sustainable development is about a better, idealized world, than what we have now. This debate is very important, as we are ultimately talking about blueprints (as humanity has done for thousands of years) of what a future society should, or should not be like.[6] The foundations for the debate in this instance, relate to the current diagnosis of the environmental and developmental state of humanity, and ideals of how progress is to be achieved from this point in time. Unlike Lomborg's view, although substantial progress has been achieved in developmental terms over the past century, the current challenges before us are much larger than he would have us believe, and importantly, that the mechanisms the international society has surrounded itself with, have the potential to worsen, not cure, these problems. The title of my recent book, *The Illusion of Progress: Unsustainable Development in International Law and Policy*,[7] encapsulates this overall thesis. Lomborg's thesis

---

[6] For the idea of progress, see Edelstein, L., 1967, *The Idea of Progress in Classical Antiquity*, John Hopkins Press, New York; Van Doren, C., 1967, *The Idea of Progress: Concepts in Western Thought*, Praeger, New York; Hilderbrand, G., 1949, *The Idea of Progress: A Collection of Readings*. University of California Press, California; Meltzer, A. (ed.), 1995, *History and the Idea of Progress*, Cornell University Press, New York; Marx, L., 1998, *Progress: Fact or Illusion*, University of Michigan Press, New York. With regard to the closely linked idea of Utopia, see Manuel, F., 1979, *Utopian Through in the Western World*, Harvard University Press, Cambridge Mass; Manuel, F. (ed.), 1965, *Utopias and Utopian Thought*, Souvenir Press, Boston; Buber, M., 1949, *Paths in Utopia*, Routledge, London; Mumford, L., 1922, *The Story of Utopias*, Viking, New York; Berneri, L., 1950, *Journey Through Utopia*, Routledge, London; Herman, A., 1997, *The Idea of Decline in Western History*, Free Press, New York.

[7] Gillespie, *The Illusion of Progress*.

is different to mine. He argues: 'things are getting better'[8] although he also acknowledges that 'things are not everywhere good, but they are better than they used to be'.[9] In many respects, I do not disagree with him on these broad conclusions. However, I do disagree on the mechanisms that caused this progress (and what we can extrapolate from it) and, perhaps more importantly, on the progress in the so-called indicators for the future, if we maintain the current international mechanisms that ultimately govern these areas in international terms.

Scepticism and the Risk to the Subject at Hand

The final point about scepticism that needs to be stressed is one which Socrates himself pointed out, in that if it is wrongly used, the overall discipline will be damaged. Thus:

There is a danger lest they should taste the dear delight [of scepticism] too early; for youngsters, as you may have observed, when they first get the taste in their mouths, argue for amusement, and are always contradicting and refuting others in imitation of those who refute them; like puppy dogs, they rejoice in pulling and tearing at all who come near them . . . . And when they have made many conquests and received defeats at the hands of many, they violently and speedily get into a way of not believing anything which they believed before, and hence, not only they, but philosophy and all that relates to it is apt to have a bad name with the rest of the world.'[10]

The risks with regard to the debate about sustainable development in this context is that the general public, rather than actually take the time to study and examine the merits of the current altercations, may take the broad conclusion of Lomborg's thesis that things are actually getting better, and disengage from the need to change accordingly. The follow on from this is that political justifications for inaction or actions in a certain, perhaps non-beneficial, direction become increasingly entrenched.

[8] Lomborg, *The Sceptical Environmentalist*, p. 3.
[9] Ibid., pp. 4, 5, 87.
[10] Jowett, B. (ed.), 1935, *The Dialogues of Plato*, Oxford University Press, Oxford, Vol. III, pp. 244–5.

In the case of Lomborg's text, this problem is magnified as he attempts to put himself across as a non-political commentator, without any specific agenda. In fact, it is arguable that this is a bit of a deception, as although Lomborg may not himself subscribe to any particular view, the majority of the conclusions from his work clearly parallel the work of many similar authors, who often reach comparable conclusions.[11] The work in question referred to here is the 1995 collection on *The True State of the Planet*,[12] and his *Earth Report 2000: Revisiting the True State of the Planet*, both edited by Ronald Bailey.[13] Lomborg's similarity to these works is not just in the subtitle of his own '*Measuring the Real State of the World*' (which is clearly used as a play on the Worldwatch Institute's annual series 'State of the World'). This is not to suggest that Lomborg has plagiarized these pieces. Lomborg's work is generally much better researched, presented, and argued than these earlier texts. However, there are clear overall

[11] With regard to specific topics which appear in both Lomborg's book and Bailey's books, similarities in conclusions can be noted. For the food conclusions, read Bailey's chapter, 'The Progress Explosion: Permanently Escaping the Malthusian Trap', in Bailey's 2000 edited collection, at pp. 1–25. With regard to the fisheries connection, see De Alessi, M., 2000, 'Fishing for Solutions: The State of the World's Fisheries', in Bailey's 2000 edition, pp. 86–114 and Jeffreys, K., 1995, 'Rescuing the Oceans', in Bailey's 1995 edition, pp. 295–338. For the climate chapter, read the contribution by Spencer, R., 2000, 'How Do We Know The Temperature of the Earth?' to the *Earth Report 2000*. For some analysis of the Kyoto Protocol, see Taylor, J., 2000, 'Soft Energy Versus Hard Facts', in Bailey's 2000 edition, pp. 116–54) (although note that Lomborg takes a more optimist view on the long-term potential of renewables) and Goklany, I., 2000, 'Richer is More Resilient: Dealing With Climate Change and More Urgent Environmental Problems', in Bailey's 2000 edition, pp. 155–7. For chemical risks, see the chapter by Safe, S., 'Endocrine Disruptors: New Toxic Menace?' in Bailey's 2000 edition, pp. 190–202.

[12] Bailey, R. (ed.), 1995, *The True State of the Planet*, McGraw Hill, New York.

[13] Bailey, R. (ed.), 2000, *Revisiting the True State of the Planet: Earth Report 2000*, McGraw Hill, New York.

similarities in many of the conclusions. In itself, this is not a problem, but, a difficulty appears when the texts are forwarded into the public debate. Here, unlike with Bailey's work, Lomborg attempts to overlook the political considerations behind his work, and suggests his work is value-free. This is not the case. This becomes problematic as those who do not deal with the text in detail obtain only part of an ultimately biased picture and accept it without recognizing the full vista with which it is connected.

Interpreting Environmental Problems

It is not my intention to demonstrate the factual mistakes in the data and its interpretation with regard to Lomborg's various chapters.[14] This task has already been done much more adequately by those intimately connected with the specific problems. However, one example will be provided as an illustration of the types of countercharges that have been laid against him. With regard to the population growth debate, Lomborg tries to dismiss the claim that population density in some countries is problematic (and that, rather, the issue is economic poverty, not overpopulation) by comparing numbers of countries with a similar national density ratio.[15] The difficulty with his analysis is that it fails to utilize the more useful and accurate indicator of density, which is how much land remains after excluding areas unsuited for human habitation or agriculture, such as deserts and inaccessible mountains. For example, according to this calculation, the population density of Egypt goes from 68 persons per square kilometre (which he highlights) if the unirrigated Egyptian deserts are excluded, to an extraordinary 2,000 people per square kilometre (which he fails to note).[16]

[14] Rennie, J., 2002, 'Misleading Math about the Earth', *Scientific American*, January, pp. 59–62. cf. B. Lomborg, 2002, 'The Sceptical Environmentalist Replies', *Scientific American*. May, pp. 9–10.

[15] Lomborg, *The Sceptical Environmentalist*, p. 45.

[16] Bongaarts, J., 2002, 'Population: Ignoring Its Impact', *Scientific American*, January, pp. 65–6.

Sustainable Development and Evolving Environmental Problems

My first substantive concern with the *Sceptical Environmentalist* is that it gives misleading snapshots of complex environmental problems. This can be seen with regard to examples of air and water pollution.

Air Pollution

One of the more recurrent assertions in Lomborg's analysis is that the air quality in a number of cities in developed countries is getting better, and in some instances, is much better than in earlier centuries. In particular: 'London air has not been as clear as it is today since the Middle Ages.'[17] He asserts that air pollution has decreased by more than 90 per cent in London since the 1930s.[18] In support of this he cites a collection of statistics, as well as the reknowned extreme smog of 1952 which killed approximately 4000 people. He goes on to suggest that around this period, air pollution was so bad, that it killed at least 64,000 extra people each year.[19] Although this assertion can be broadly supported as a historical fact[20] (although the 64,000

[17] Lomborg, *The Sceptical Environmentalist*, p. 164.
[18] Ibid., p. 11.
[19] Ibid.

[20] In 1880, 2200 Londoners died in a single incident when coal smoke from home heating and industry combined to form a lethal toxic smog. A further 500 died in a similar situation in 1873. The next notable case was in 1951, when the (very cold and stable) weather conditions in early winter trapped the pollutants over the capital. The smog extended for 30 km around London, and visibility was reduced to 1–5 metres. The pH level of the air was between 1.6 and 2—roughly the equivalent of the chemical make-up of sulphuric acid (ten times what the WHO currently recommends as safe). This smog was linked to the deaths of 1850 people. The following year, at the same time and under similar conditions, two further smogs killed 4700 people. In the London winter of 1962, 750 more people were directly

figures needs to be checked), the assertion that he goes on to extrapolate—that air pollution has declined by a radical extent overall, can be disagreed with. Rather, what would be more accurate to suggest is that the type of air pollution that Lomborg implicitly fingers from an historical perspective—sulphur dioxide, from large-scale power stations and domestic heating—has fallen dramatically. However, to extrapolate from this that air pollution in general has fallen in London (or other major cities in the developed world) is mistaken.

That is, although sulphur pollution from traditional sources has clearly dissipated,[21] and moderate reductions have been made with nitrogen oxides (but hardly the type of vast reductions he suggests),[22] the overall problem of air pollution has not disappeared. Rather, one source of air pollution has been largely replaced by another, which is typically manifested as very small

---

killed. Similar results were recorded in the Meuse Valley in Belgium in 1930 (63 deaths) and Donoara in Pennsylvania in 1948 (28 deaths, 14,000 ill). For a discussion of these issues, see World Resources Institutes, UNEP and World Bank, 1999, *World Resources: 1998–1999*, Oxford University Press, Oxford, p. 63; McCormick, J., 1997, *Acid Earth*, (third edition, Earthscan, London, pp. 5, 32. See Pearce, F., 1992, 'Back to the Days of Deadly Smogs,' *New Scientist*, December 5, pp. 25–6, R. Read, 1991, 'Breathing Can Be Hazardous to Your Health', *New Scientist*, 23 February, pp. 26–9; Hamer, M., 1984, 'Ministers Opposed Action on Smog', *New Scientist*, 5 Jan, p. 3; Anon, 1986, 'Pea Soupers and Westland Helicopters in 1955', *New Scientist*, 2 Jan, p. 12.

[21] Taken as a whole, the 21 Parties of the 1985 Helsinki Protocol on the Reduction of Sulphur Emissions reduced 1980 sulphur emissions by more than 50 per cent by 1993. In the whole of Europe, including non-Parties to the Protocol, the sum of emissions was 43 per cent lower than in 1980. Four non-parties achieved a 30 per cent reduction (or more), 11 Parties achieved reductions of at least 60 per cent and two of these had reductions above 80 per cent. By 1994, a 50 per cent average reduction for the 21 Parties had been achieved.

[22] Holdren, J., 2002, 'Energy: Asking the Wrong Questions', *Scientific American*, January, pp. 63–5.

airborne particles. Airborne particles are referred to as suspended particulate matter (SPM), total suspended particulates (TSP), or black smoke, depending on the type of measurement used.[23] Particulates in the atmosphere may be divided into two principal size groups, fine particles (up to 2 microns in diameter) which come from combustion processes and from coagulation and condensation of gases and vapours, and larger particles (coarse particles 2 to 100 microns in diameter). It is the smaller sized particles that represent the greatest danger to human health.[24] Moreover, because fine particulate matters (PMs) are typically by-products of combustion particles, they are more likely to contain carcinogens.[25] Many of these fine particles penetrate deep into the lungs and cause inflammation, resulting in the release of molecules called cytokines. These could, in turn, trigger changes in the heart's blood vessels.[26] Like traditional air pollutants, they can kill vast numbers of people (although unlike the episodic events that Lomborg cited, SPM pollution tends to be more slowly acting; however, it eventually reaches the same conclusion). For example, in the United States, death rates increase in almost direct proportion to the level of particulate pollution. Life expectancy may be 2–3 years shorter in communities with high PM than in communities with low PM.[27] Much of this

---

[23] Whatever the name, the source is the same. Thus, when coal and certain other fuels burn, they emit substances, including carbon particles (if combustion is inefficient) and SO_2 gas. In addition, the high temperature of combustion cause nitrogen in the air to combine with oxygen, yielding nitrogen oxide (NOx) gases. Shaw, R., 1987, 'Air Pollution by Particles', *Scientific American*, August, p. 84.

[24] Anon., 1993, 'Deadly Urban Air', *New Scientist*, 29 May, p. 11; Hamer, M., 1996, 'Clean Air Strategy Fails to Tackle Traffic', *New Scientist*, 31 August, p. 6.

[25] Boyce, N., 2000, 'Hold Your Breath', *New Scientist*, 5 August, p. 5.

[26] Day, M., 1998, 'Taken to Heart', *New Scientist*, 9 May, p. 23.

[27] People living in the most polluted city—Steubenville, Ohio—had a 26 per cent risk of dying young compared with residents of the cleanest city (Portage, Wisconsin). WHO, 1999, *Protection of the Human Environment:*

damage appears to begin at a young age. For example, children's DNA can be damaged by compounds called polycylic aromatic hydrocarbons which are produced when fuel burns and coat PM10s (particulate matter upto 10 microns in diameter). Some of this damage is done when the foetus is in the womb.[28] In the early 1990s, PM10s were considered to be killing up to hundreds in Paris,[29] thousands in the UK,[30] and tens of thousands in Europe. Research in New Zealand suggests that nearly an equal number of people die annually from SPM related illnesses (399 people over 30 years of age die prematurely), as are killed in vehicle accidents (454).[31]

That SPM pollution is a problem is not news to Lomborg. Rather, he accepts that SPM pollution is a distinct (relatively new) problem.[32] My concern here is not that he argues that particle pollution is being increasingly controlled in developed countries[33]

Air Quality Guidelines, WHO, Geneva, vol. 3, p. 39. Boyce, N., 2000, 'Hold Your Breath', *New Scientist*, 5 August, p. 5; UNEP, 2000, GEO 2000 Report, Earthscan, London, p. 168.

[28] Edwards, R., 1996, 'Smog Blights Babies in the Womb', *New Scientist*, 19 October, p. 8.

[29] Patel, T., 1996. 'French Smog Smothers Hundreds', *New Scientist*, 17 February, p. 7.

[30] The original figure was 10,000 people in the UK each year. Hamer, M., 1994, 'Dying from Too Much Dust', *New Scientist*, 12 March, p. 5. In 1998, this figure was lowered to 8100 per year. See Day, M., 1998, 'City Dwellers Dying for a Breath of Fresh Air', *New Scientist*, 24 January, p. 16. Cf. Hamer, M., 1997, 'Lies, Damned Lies', *New Scientist*, 29 November; M. Hamer, 1996, 'Cars Must Go to Meet Clean Air Targets', *New Scientist*, 18 May, p. 12; Anon., 1999, 'Not So Clean', *New Scientist*, 23 January, p. 5; Editor, 1994, 'Smog Alert', *New Scientist*, 25 June, p. 3. Editor, 1995, 'Britain's Last Gasp', *New Scientist*, 13 May, p. 3.

[31] New Zealand Press Association, 2002, 'Vehicle Pollution Major Killer', *New Zealand Herald*, 22 March, p. A8.

[32] 'It is only within the last decade that we have realised how dangerous airborne particles actually are.' Lomborg, p. 167.

[33] Lomborg, *The Sceptical Environment*, p. 169.

(he overstates his case)[34] but his failure to recognize the contradiction in his overall position that air pollution in developed countries is rapidly declining. Indeed, Lomborg notes that particle pollution is estimated to cause roughly 135,000 premature deaths in the USA each year—or almost 6 per cent of all deaths. In the UK, a similar number of deaths is estimated at 64,000.[35] The interesting thing about this 64,000 figure is that it is exactly the same as the historical example figure of deaths he cites, from which apparently there has been a massive decline. Either there is a mistake in the text, or Lomborg has failed to point out a key concern (and this applies to many areas of sustainable development in general)—although elements of one section of an overall environmental problem may have been addressed, often they have been replaced by equally vexing problems which have emerged with a changing society under the same broad banner. In this instance, the problem of taking away traditional sources of air pollution (large-scale coal-fired power plants and domestic heating with inefficient fuels), has largely been replaced by other sources— such as the transportation sector, or more particularly, exhausts from motor vehicles, and diesel exhausts in particular.[36]

The other area with regard to air pollution in which Lomborg makes a similar mistake is not with the sources of air pollution, but with regard to its impact upon nature. In particular, Lomborg seizes upon the 1980s debate, in which acid rain (as it was initially known) was supposed to be having a detrimental

---

[34] Typical limits for SPM 2.5s in Los Angeles are 20 micrograms and 16 micrograms in New York. The limit set by the Environmental Protection Agency is 15 micrograms. Although British levels are similar, sites in London record averages of up to 32 micrograms. Pearce, F., 2002, 'Big City Killer' *New Scientist*, 9 March, p. 8.

[35] Lomborg, *The Sceptical Environmentalist*, p. 168.

[36] World Resources Institute, UNEP and World Bank, 1999, *World Resources: 1998–1999*, Oxford University Press, Oxford, p. 63; UNEP/ WHO, 1994, 'Air Pollution in the World's Megacities', *Environment*, vol. 36, no. 2, p. 11; Pearce, F., 1999, 'Burning Rubber', *New Scientist*, 10 April, p. 14.

impact on forests. Lomborg discounts this, suggesting that while acid rain did harm lakes it 'did very little if any damage to forests'.[37] Although he accepts that damaged trees exist in Europe, he suggests that: 'in reality [such a problem is] simply a non-specific expression that applies to numerous specific, familiar diseases, and the reason why we have started worrying about it is that we have started monitoring this loss'.[38] However, rather than downplaying such a threat (as Lomborg's analysis would suggest), in 2000 the European Commission report on the State of Europe's Trees suggested that two-thirds of the trees were sickly, and only 36 per cent of all broadleaf and conifers are healthy. One in five European trees show signs of damage, having lost at least a quarter of their leaf canopy. The rest are visibly dropping more than usual.[39] At this point the question becomes: what is causing the damage? The answer is probably a cocktail of pollutants, which reveals a much more complex problem than when it emerged in the 1980s. As such, Lomborg is correct in that a linear causal relationship between acid rain and its effect on forests (as traditionally argued and understood) has not come to fruition. However, this is not to say that the problem has gone away. Rather, the nature of the impact of the problem, due to an evolving understanding of cumulative contributions of diverse pollutants (i.e., damage is linked not only to acidification, but also to critical loads of nitrogen and low-level ozone) suggests a much more scientifically complicated situation.

When viewed from this broader perspective, it is possible to examine not only the effects upon forests, but upon ecosystems in general. Lomborg largely omits this broader picture. A good example is the eutrophication of ecosystems, where although

[37] Lomborg, *The Sceptical Environmentalists*, p. 13.

[38] Ibid., p. 180. For a scathing commentary on this analysis, see Lovejoy, T., 2002, 'Biodiversity: Dismissing Scientific Processes', *Scientific American*, January, pp. 67–9.

[39] Jones, N., 2000, 'Crisis Time for Europe's Ravaged Forests', *New Scientist*, 28 October, p. 6.

progress is being made (and especially due to the strong international air pollution regime and its long term targets)[40] as of 1998, in central Europe more than 90 per cent of ecosystem areas were exposed to critical loads of pollutants which cause eutrophication. This situation is widespread. In 70 per cent of the countries where the Long Range Transboundary Air Pollution convention and Protocols apply, more than half of the ecosystem areas are affected by eutrophication. The situation has shown little improvement since 1990. Moreover, the percentage of ecosystem areas exposed to eutrophication has increased since 1990 in about half of the countries in the area.[41]

Ocean Pollution

A second example of Lomborg presenting a distorting picture relates to pollution of the oceans. To prove his thesis that the health of the oceans is actually much better than suggested, he utilizes accidents from oil tankers, eutrophication of coastal waters, and pollutant levels in shellfish and fish. With regard to declining accidental oil spills,[42] and the threats to the oceans posed by oil spills, he is broadly correct (that the amount entering the ocean from this source has declined).[43] However,

---

[40] Once the Gothenburg Protocol is implemented, the area in Europe with excessive levels of acidification should shrink from 93 million hectares in 1990 to 15 million hectares in 2010. Excessive levels of eutrophication should also fall from 165 million hectares in 1990 to 108 million hectares in 2010. UN/ECE (2000), 'The 1999 Gothenburg Protocol to Abate Acidification, Eutrophication and Ground Level Ozone', http://www.unece.org/env/lrtap/multi_h1.htm; Wettestad, J., 1997, 'Acid Lessons? LRTAP Implementation and Effectiveness', *Global Environmental Change*, vol. 7, no. 3, pp. 235–49.

[41] EMEP, 2000, *Transboundary Acidification and Eutrophication in Europe*, EMEP, Geneva, p. 11.

[42] Lomborg, *The Sceptical Environmentalist*, p. 191.

[43] This improvement has been due to increasingly safe vessel designs ('1991 MARPOL Amendments Enter into Force', *IMO News*, vol. 2, 1993, p. 2. This ideal eventually appeared as Regulation 25A in Annex 1.

what he fails to point out is that although (as of 2002)[44] tanker and pipeline spills put 150,000 tonnes of oil into the ocean (well down from this historical percentage contributions) a further 38,000 tonnes come from oil well leaks. However, the lion's share of oil pollution in the ocean (480,000 tonnes) comes from run-off from land spills and emissions from small boats used for recreation.[45] As such, once more, Lomborg has given a false impression that the environmental threat has been dealt with. Although this is partly correct in that one original source has been confronted, in reality, a different problem remains as the source has changed. Moreover, in this instance, the new source is not being well confronted, as unlike international tanker accidents, which fall squarely within a well formed body

See also 1994 Year Book of International Environmental Law, Year Book of International Environmental Law. vol. 5, pp. 184–5, enhanced international and regional co-operation in dealing with oil accidents at sea. This process began with the 1969 International Convention Relating to Intervention on the High Sea in Cases of Oil Pollution Casualties. For the Regional Agreements on Oil Pollution, see the 1969 Agreement for the North Sea; the 1976 Protocol Concerning the Mediterranean; the 1978 (Kuwait) Protocol; the 1981 (Abidjan) Protocol; the 1981 (Lima) Agreement, the 1981 (Jeddah) Protocol; the 1983 Protocol on the Wider Caribbean Region; the 1983 Agreement for the North Sea; the 1986 Protocol for the Eastern African Region; the 1986 Protocol for the South Pacific Region; and strong supplementing international liability regimes. See for example the 1969 Convention on Civil Liability for Oil Pollution Damage; 973 UNTS. 3; 1971 Convention on the Establishment Fund for Compensation for Oil Pollution Damage, 11 ILM, (1971), 284; International Convention on Liability and Compensation for Damage in Connection With the Carriage of Hazardous and Noxious Substances by Sea (1996), 35 ILM, 1406).

[44] Lomborg's foundation citation in this area is from 1985. Lomborg, *The Sceptical Environmentalist*, p. 189.

[45] Anon., 2002, 'Dead in the Water', *New Scientist*, 1 June, p. 7; Group of Experts on the Scientific Aspects of Marine Pollution (GESAMP), 1990, *The State of the Marine Environment*, UNEP, Nairobi.

of international law, pollution from land-based sources are not.[46]

Another example of utilizing deceptive statistics deals with his assertion that with regard to pollution in coastal waters from heavy metals and persistent organic pollutants, studies in Denmark and the United Kingdom, show that: 'concentrations of harmful substances such as DDT (Dichloro diphenyl trichloro ethane), PCB (Polyclorinated biphenyls), dieldrin, and cadmium have fallen drastically in coastal seas'.[47] Although this may be correct for the situation he cites (and he chooses concentrations in shellfish and fish), a different outcome may be obtained if looking at marine mammals (typically small cetaceans, which are tragically good species at bioaccumulation of pollutants).[48] This problem has been detected mostly around key Northern countries[49] although it is also evident in the waters surrounding tropical countries and in the Southern hemisphere.[50]

[46] See Gillespie, A., 2002, 'Environmental Threats to Cetaceans and the Limits of Existing Management Structures', 6 *NZ Journal of Environmental Law*, 97–139.

[47] Lomborg, *The Sceptical Environmentalist*, p. 195.

[48] Reijnders, P. and Simmonds M. (ed.), 'Report of the Workshop on Chemical Pollution and Cetaceans', *Journal of Cetacean Research and Management: Chemical Pollutants and Cetaceans, Special Issue*, vol. 1, pp. 3–14; Parsons, E., 1999. 'Immune Level Abnormalities', *Veterinary Record*, vol. 144, pp. 75–6; Troisi, T., 1996, 'Toxic Effects of PCBs on Marine Mammals', *Soundings*, vol. 2, no. 8, pp. 1–2; Colborn, T., 1996, 'Epidemiological Analysis of Persistent Organic Pollutants in Cetaceans', *Reviews of Environmental Contamination and Toxicology*, vol. 146, pp. 91–172; Edwards, R., 1999, 'Sea Sickness: Deaths of Harbour Porpoises are Linked to PCBs and Mercury', *New Scientist*, 18 December, p. 12; Beland, P., 1996, 'The Beluga Whales of the St. Lawrence River,' *Scientific American*, May, pp. 58–65.

[49] Simmonds M., 2000, 'Cetacean Contaminant Burdens: Regional Examples', SC/51/E 13; Report of the Scientific Committee, IWC/53/4, p. 61.

[50] See Parsons, E., 2002, 'The Impact of Pollution on Humpback Dolphins,' SC/54/SM5; Ministry for the Environment, 1997, *The State of New Zealand's Environment*, MFE, Wellington, vol. 9, p. 132.

With such considerations in mind, a problem arises when it becomes apparent that in certain places, people are consuming cetaceans (which represent a very strong pathway for POPs to enter into the human body) that contain concentrations of POPs that continue to accumulate and bio-magnify up the food chain to dangerous levels. For example, in the Arctic where POP contamination can be 10–20 times higher than in most temperate regions, indigenous people who rely on traditional diets are likely to be more exposed to several toxic substances than the majority of people elsewhere in the world. Along the west coast of Greenland, in Nunavik, Canada, and in Nikel on the Kola Peninsular of Russia, blood levels of, *inter alia*, DDT are only a small fraction lower than the levels that are known to have caused neurological defects in babies. Detrimental PCB contamination is a strong problem at Nunavik, and in northwest Greenland, where fetal and childhood development may be at direct risk. This conclusion is not surprising in Greenland, where more beluga and narwhal is consumed than anywhere else, as 95 per cent of women exceed the Canadian guideline limit for PCB contamination (five parts per million). Such problems, which have been associated with the consumption of POP infested products (typically cetaceans), have not been restricted to the Arctic.[51]

A similar body of evidence is also now developing in Japan in regard to general toxicity problems arising from seafood (in general, and whale meat products in particular) which may contain several sorts of environmental toxins at levels above the safety limits prescribed by most national and international authorities. For example, 2002 samples from the Tokyo market revealed dolphin meat which contained 2000 micrograms of mercury per gram. This is 5000 times higher than the 0.4

---

[51] Arctic Monitoring and Assessment Programme, (AMAP) 1997, *Arctic Pollution Issues: A State of the Arctic Environment Report*, AMAP, Oslo, pp. 3.28, 5.1, 5.6, 5.7, and 5.16; Swiss Coalition for the Protection of Whales/ Global Survival Network (GSN), 1999, *Polar Exposure: Environmental Threats to Arctic Marine Life and Communities*, GSN, London, p. 16.

microgram safety level.[52] This situation has led to a series of resolutions from the International Whaling Commission (IWC), and a number of sovereign countries, recommending limits on how much cetacean products should be consumed.[53]

Sustainable Development, Risk, and the Precautionary Approach

Lomborg clearly dislikes the idea of any 'catastrophe' or 'environmental surprise' landing on the doorstep of humanity. His argument against this can be seen via his discussions about climate change, and later about the precautionary principle. With regard to climate change, the basis of Lomborg's position derives from the scenarios offered by the Intergovernmental Panel on Climate Change (IPCC), which has moved away from its previous 'best guess' scenarios for future emissions. Rather, the IPCC has prepared a number of future emission scenarios. Each of the future predictions is highly dependent on demographic change, social and economic development, and the rate and direction of technological change. Within each one of these sectors, dozens of inter-relating issues overlap. The result is a series of future emission scenarios which range, for 2100, from 770 GtC to 2540 GtC. There is no single most likely or 'best guess' scenario with the current scenario literature, as far too many uncertainties and variables make this impossible. Thus, each option is equally possible, depending on how humanity responds to the various considerations.[54] The consequence is that

[52] Anon., 2002, 'Bad Catch', *New Scientist*, 23 March, p. 5; Coghlan, A., 2002, 'Its Madness', *New Scientist*, 1 June, p. 17.

[53] See IWC Resolution 2000–6, 'Resolution on POPs and Heavy Metals; IWC, 52nd Report, 2001, p. 67; See Appendix 5, IWC Resolution 1999–4, Resolution on Health Effects from the Consumption of Cetaceans; see IWC, 50th Report, 1999, p. 53.

[54] IPCC, 2000, *Emission Scenarios*, Cambridge University Press, Cambridge, pp. 1–15.

the higher the build-up of greenhouse gases, the worse the knock-on effects of enhanced climatic change.

Unlike the IPCC, which adopts the classical point that Socrates did, and suggested that certainty in this area is not possible due to so many uncertainties, Lomborg adopts the opposite position. He argues that all but the lower scenarios for temperature increases are largely unrealistic, and subsequently tends to overestimate the speed of global warming. As such, Lomborg suggests that a more realistic basis to work from is the lesser scenarios.[55] Accordingly, he surmises that many of the problems with climate change will be of a much lesser magnitude and subsequent impact, and as such, should be ultimately downgraded as a serious risk facing humanity (especially in comparison to other problems).[56] He goes so far as to suggest that climate change will produce overall positive benefits, such as general increases in global biomass, due to the so-called fertilization effect,[57] and lengthened growing seasons for agriculture[58] (although he conceded there may be some effects on developing countries, but these should be offset by overall increases elsewhere).[59] His conclusion of the overall output of agriculture is broadly in congruence with accepted analysis in this area,[60] but the uncertainty in this area is high. This is especially so when additional macro-economic considerations, like future trade flows in food stuffs are added,[61] in addition to the vulnerability of individual agricultural

[55] Lomborg, *The Sceptical Environmentalist*, p. 287.

[56] Ibid., pp. 318, 322, 323.

[57] Ibid., p. 299.

[58] Ibid., p. 28.

[59] Ibid., pp. 288–9.

[60] IPCC, 1996, *Climate Change 1995: Impacts, Adaptations and Mitigation*, Cambridge University Press, Cambridge, p. 9.

[61] IPCC, 2001, *Climate Change 2001: Impacts, Adaptation and Vulnerability*, Cambridge University Press, Cambridge, p. 9; Reilly, J., 1994, 'Climate Change and Agricultural Trade', *Global Environmental Change*, vol. 4, no. 1, pp. 24–36.

sectors (i.e. depending on how adverse to risk they are).[62] Finally, and perhaps most importantly, all of the scenarios are dependent on the actual climate changes. The more serious the changes, the greater the impacts upon agriculture that may be expected, and rather than any form of positive increases, negative impacts[63] may predominate overall. The negative impacts may be concentrated in certain regions.[64] For example, in 2001, UNEP suggested that harvests of vital crops like rice, wheat, and corn could plummet by over a third (a 10 per cent drop for every 1 centigrade increase) in some regions over the next one hundred years, causing mass starvation in some regions.[65]

The important point of the above discussion is that uncertainty is rife in this area—as with other areas related to sustainable development. Moreover, the uncertainties of impact may become magnified with additional uncertainties, such as environmental surprise, being added into the equation.

With regard to climate change, it is possible that the Earth may respond in unanticipated ways to forced climate change. In

[62] IPCC, 1998, *The Regional Impacts of Climate Change: An Assessment of Vulnerability*, Cambridge University Press, Cambridge, p. 6.

[63] IPCC, *Climate Change 2001*, p. 4; Joyce, C., 1991, 'World's Rice Crop Vulnerable to Changing Atmosphere', *New Scientist*, 12 January, p. 16; Nowak, R., 2002, 'How the Rich Stole the Rain', *New Scientist*, 15 June, p. 6; Mulvaney, K., 1998, 'Eaten Alive', *New Scientist*, 18 July, p. 12; Holmes, B., 1998, 'Unwelcome Guests', *New Scientist*, 18 April, p. 15; Copley, J., 1999, 'Off Their Food', *New Scientist*, 3 April, p. 23; Kleiner, K., 1998, 'Dying for a Change', *New Scientist*, 5 September, p. 8; Pain, S., 1988, 'No Escape from the Greenhouse', *New Scientist*, 12 November, p. 38; Bazzar, F., 1992, 'Plant Life in a CO_2 Rich World', *Scientific American*, January, pp. 18–24; Gribbin, J., 1995, 'Plants Issue their Own Global Warming,' *New Scientist*, 30 September, p. 21; Fajer, E., 1992, 'Is Carbon Dioxide a Good Gas?', *Global Environmental Change*, pp. 301–10.

[64] IPCC, *Climate Change 2001*, p. 4.

[65] Pearce, F., 2001, 'Global Warning', *New Scientist*, 17 November, p. 4; Pearce, F., 1992, 'Grain Yields Tumble in Greenhouse World', *New Scientist*, 18 April, p. 4; Rind, D., 1995, 'Drying Out', *New Scientist*, 6 May, pp. 36–41.

the literature on climate change this is known as 'surprise'.[66] Within the official literature, the IPCC warned in 1990 that despite their predictions: 'The complexity of the system means that we cannot rule out surprises'.[67] The IPCC 1996 report also emphasized the possibility of 'surprises and unanticipated rapid change'.[68] The Third Assessment Report in 2001 by the IPCC added that the potential for large-scale, and possibly irreversible impacts, poses risks that have yet to be reliably quantified. These possibilities are very climate scenario-dependent and a full range of plausible scenarios has not yet been evaluated. Examples include significant slowing of the ocean circulation that transports warm water to the North Atlantic, large reductions in the Greenland and West Antarctic Ice Sheets, accelerated global warming due to carbon cycle feedbacks in the terrestrial biosphere, and releases of terrestrial carbon from permafrost regions and methane from hydrates in coastal sediments.[69] These risks may be more pronounced if the carbon more than doubles (above pre-industrial levels) in the longer term.[70]

The obvious way to factor in the extreme 'surprise' events, in addition to the uncertainties with regard to the large-scale risks that climate change represents, is to proceed with a precautionary principle/approach and seriously contemplate strong climate mitigation policies (as the responsible scientific community cannot rule out such catastrophic outcomes at a high level of

---

[66] Streets, D., 2000, 'Exploring the Concept of Climate Surprise,' *Global Environmental Change*, vol. 10, 97–107; Pain, S., 1992, 'Runaway Greenhouse Warming Cannot Be Ruled Out', *New Scientist*, 15 February, p. 19; Leggett, J., 1992, 'Running Down to Rio', *New Scientist*, 2 May, pp. 37–42.

[67] IPCC, noted in Milne, R., 1990, 'Pressure Grows on Bush to Act on Global Warming', *New Scientist*, 2 June, p. 5.

[68] IPCC, noted in Pearce, F., 1995, 'Global Warming Jury Delivers Guilty Verdict', *New Scientist*, 9 December, p. 6.

[69] IPCC, *Climate Change 2001*, p. 7.

[70] IPCC, *Climate Change 1995*, p. 4.

confidence).[71] However, Lomborg argues against the precautionary principle being applied in this area. He did this because he believed that everyday society deals with issues which may have impacts on it of a negative or positive nature (he cites the advent of the Internet affecting personal contacts, though superfluous, as an example) but it does not apply (nor need to apply) the precautionary principle. He then goes on to extrapolate that:

the environmental area has been able to monopolise the precautionary principle is in essence due to the Litany and our fear of doomsday. Of course, if large scale ecological catastrophes were looming on the horizon we might be more inclined to afford the extra margin of safety just for the environment. But as is documented in this book, such a general conception is based on a myth.[72]

At this point, Lomborg makes the mistake of failing to know the history of one of the 'large scale ecological catastrophes' (which he only very briefly examines in his text)[73]—that of ozone depletion. The point of this example is that following the proposal of the scientific hypothesis of ozone depletion in 1974, there was not actual 'proof' of ozone depletion until over a decade later. During this time, the theoretical risks of ozone depleting substances were continually downgraded (for a number of reasons). However, when it was eventually discovered (the thinning ozone layer over Antarctica), it was shown that nature was responding in completely unanticipated ways, and as such 'surprised' humanity. In particular, rather than providing a neutral sink for the chlorine based chemicals to safely nestle,[74]

---

[71] Schneider, S., 2002, 'Global Warming: Neglecting the Complexities', *Scientific American*, January, pp. 61–3.

[72] Lomborg, *The Sceptical Environmentalist*, p. 350.

[73] Ibid., pp. 273–4.

[74] Gribbin, J., 1979, 'Monitoring Halocarbons in the Atmosphere', *New Scientist*, 18 January, pp. 164–7; Eggleton, A., 1976, 'Will Chlorofluorocarbons Really Affect The Ozone Shield?', *New Scientist*, 20 May, pp. 402–3; Anon., 1976, 'Upper Atmosphere Chemistry: The Arguments Continue', *New Scientist*, 10 June, p. 564; Anon., 1976, 'Aerosols and Ozone: Good News and Bad', *New Scientist*, 20 May, p. 395.

it was shown that when the chemicals finally accumulated at the Poles, rather than being beneficial, the special circumstances of the area catalysed the situation into being very detrimental. As such, the unique natural conditions, more than being some form of excuse for the ozone depletion (as initially suggested),[75] provided an example of where nature behaved in a completely unpredictable way and exacerbated an anthropogenically caused problem.[76] As if this omission of prediction was not enough in itself, predictions that a similar destruction would not occur in the Arctic due to warmer temperatures and less stable air masses, and different chemical mixes, were repeatedly shown to be wrong in the 1990s.[77] As such, the thinning of the ozone layers over

[75] Anon., 1986, 'Ozone Hole is Normal,' *New Scientist*, 4 September, p. 22.

[76] Specifically, the particular uplifting of air currents after the dark Antarctic winter, whereby ozone-poor air from lower altitudes is swept into the stratosphere via a 'polar vortex' culminating with extremely cold temperatures and stable air masses, to form the perfect platform for ozone depleting substance destruction of the ozone layer. This isolation allows temperatures to fall to the point where icy polar stratospheric clouds form. The first spring sunlight triggers photochemical reactions which release active chlorine from otherwise inert compounds on the surface of ice crystals in the clouds. This contains chlorine that destroys the ozone layer. Later in the spring, warm air dissolves the vortex and the stratosphere returns to more normal conditions. Anon., 1987, 'US Spy Plane Set to Examine Origins of Ozone Hole', *New Scientist*, 30 July, p. 22, Gribbin, J., 1987, 'An Atmosphere in Convulsions', *New Scientist*, 26 November, pp. 30–1; Stolarski, R., 1988, 'The Antarctic Ozone Hole', *Scientific American*, January, pp. 20–5; Verma, S., 1989, 'As Antactica's Ozone Hole Grows,' *New Scientist*, 7 October, p. 9; Joyce, C., 1987, 'Chlorine Clears the Ozone Layer Down South,' *New Scientist*, 8 October, pp. 18–19; Anon., 1987, 'Lab Experiments Back Theories of Ozone Depletion', *New Scientist*, 3 December, p. 28.

[77] Final Report: Second Session of the First Meeting of the Open Ended Working Group of the Parties to the Montreal Protocol, UNEP/OzL.Pro.WG.1(2)/4, p. 15; First Meeting of the Parties to the Montreal Protocol, Helsinki, 2–5 May. UNEP/OzL.Pro.1/5, 6 May 1989, Paragraph

the Arctic and the Antarctic was the first time that non-linear responses became apparent in international environmental issues. That is, a 'runaway effect', which was not only completely unexpected,[78] but has also managed to confound predictions, and has provided a large part of the justification for the international community to act, in the future, in a precautionary manner when dealing with such issues.

Sustainable Development and the Technology Question

Lomborg is clearly a technological optimist (for certain, but not all, technologies)[79] when it comes to environmental problems.[80] Indeed, with many of the problems he cites as solved, he suggests that, in essence, they have been bypassed by simple application of newer, more efficient technologies. This analysis applies to food shortages and the Green Revolution,[81] the ozone problem,[82] and

14; Toon and Turco, R., 1991, 'Polar Stratospheric Clouds and Ozone Depletion', *Scientific American*, June, p. 40–7; Pearce, F., 1997, 'Nature Fuels Loss of Arctic Ozone,' *New Scientist*, 7 June, p. 11; Editor, 1996, 'Disaster in the Stratosphere', *New Scientist*, 16 March, p. 3; Anon., 1996, 'Hole Over Britain', *New Scientist*, 16 November, p. 11; Hecht, J., 1999, 'Polar Alert,' *New Scientist*, 12 June, p. 6; Editor, 1987, 'The Ozone Zone', *New Scientist*, 12 November, p. 8; Farman, J., 1987, 'What Hope for the Ozone Layer Now?', *New Scientist*, 12 November, pp. 50–4; MacKenzie, D., 1988, 'Coming Soon: The Next Ozone Hole', *New Scientist*, 1 September, p. 38.

[78] Gribbin, J., 1987, 'An Atmosphere In Convulsions', *New Scientist*, 26 November, p. 30–1; Anon., 1988, 'Farman Calls for Tighter Controls on CFCs', *New Scientist*, 24 March, p. 23.

[79] Lomborg casts a sceptical eye over the possibilities of modern renewable technologies, such as wind and solar over the short term, Lomborg, *The Sceptical Environmental*, p. 19.

[80] 'Technology makes it possible to achieve growth as well as a better environment . . . ., Ibid., p. 176.

[81] Ibid., p. 62–5.

[82] 'It is worth pointing out that the implementation of the CFC ban was strictly profitable. It was actually relatively cheap to find substitutes for CFC (e.g. in refrigerators and spray cans).' Ibid., p. 274.

air pollution. With the last problem, he suggests sulphur dioxide pollution has declined because of, *inter alia* (in a historical context): 'a move away from siting power plants in urban areas and the use of taller smoke stacks'; and in a contemporary context: 'cars pollute much less than they used to, partly because of catalytic converters but also because diesel vehicles now use low-sulphur fuels'.[83] Finally, with regard to mitigating climate change he suggests:

> We should be much more open towards other techno-fixes [as opposed to just new renewables such as wind and solar power]. These suggestions range from fertilising the ocean (making more algae bind carbon when they die and fall to the ocean floor) and putting sulphur particles into the stratosphere (cooling the Earth) to capturing carbon dioxide from fossil fuel use and returning it to storage in geological formations.[84]

The difficulty with this type of analysis is that although sometimes technology can quickly solve environmental problems and has a central role in sustainable development,[85] in a number of other situations, the new technologies that are designed to solve one problem may quickly create another. For example, with the ozone situation, it is true that the original ozone depleting substances (typically chlorofluorocarbons) were replaced by hydrochlorofluoro-carbons (HCFCs) and hydrofluorocarbons (HFCs).[86] However, these replacements have not been clear-cut. That is, although HCFCs have a lesser impact upon the ozone layer than their predecessors, they still have a detrimental impact (and have been subsequently restricted).[87] Likewise, with regard to HFCs, although they have no effective ozone depleting

---

[83] Ibid., pp. 169–70.

[84] Ibid., p. 323.

[85] See UNDP, 2001, *Human Development Report 2001: Making New Technologies Work for Human Development*, Oxford University Press, Oxford.

[86] Report of the 11th MOP, UNEP/OzL.11/10, 7 December 1999, p. 8.

[87] Anon., 1990, 'Scientists Warn of Perils Posed by Substitutes for CFCs', *New Scientist*, 30 June, p. 7.

potential, they are listed chemicals under the Kyoto Protocol due to their extreme global warming potential.[88] A second example of this type of problem deals with Lomborg's comment on 'tall stacks' as a way of (historically) reducing air pollution. Lomborg is correct on this ground, and the building of tall stacks as a measure to defeat local air pollution was actively pursued in a number of countries after 1950. For example, in 1955 only 2 stacks in the US were taller than 180 metres. By 1980, all stacks being built were taller than that. By 1975, at least 15 stacks in the world were taller than 300 metres. Although this option was cautiously supported by some early international policy suggestions in this area, the support for this policy disappeared when it became apparent that these tall stacks merely turned a problem of local air pollution to one of transboundary air pollution. Thus, by the early 1980s, the 'tall chimney principle' was abandoned at an international conference, as the transferring of pollution from local to regional neighbours was deemed unacceptable.[89]

Working from Lomborg's list, a third example of this problem can be seen in his reference to diesel fuel. While it is possible to control aspects of diesel pollution, it is important to realize the original source of this current problem. That is, diesel engines were initially enthusiastically embraced as an alternative to conventional engines. This was because they produce only about 10 per cent as much carbon monoxide and hydrocarbons as conventional vehicles fitted with catalytic converters. Emissions of nitrogen oxides are also thought to be two-thirds less than in conventional cars, and their emissions of volatile organic compounds are also much reduced. Moreover,

---

[88] Decision X/16. Implementation of the Montreal Protocol in Light of the Kyoto Protocol. Report of the 10th MOP of the Montreal Protocol, UNEP/OzL.Pro10/9, 3 December, pp. 30–1.

[89] Swedish Ministry of Agriculture, 1982, The 1982 Stockholm Conference on the Acidification of the Environment, Stockholm, p. 19.

they burn 25 per cent less fuel, and the fuel is typically cheaper than petroleum.[90] Despite these advantages, it was later revealed that diesel vehicles emit between 10–100 times more SPMs than conventional cars.[91] In the United States, diesel engines account for up to 70 per cent of smoke emissions in some areas. In the UK (as Lomborg himself notes) although diesel cars make up only 6 per cent of the total vehicles, they contribute with 92 per cent of all vehicle emissions.[92] Aside from the recognized SPM threats, in the late 1990s a compound discovered in the exhaust fumes of diesel engines, which may add considerably to the mutagenic activities of particles, is 3-nitrobenzanthrone—possibly the most carcinogenic ever analysed from vehicle exhausts.[93] Lomborg's retort to this collection of problems is that it is technologically possible to capture all of the pollutants that diesel engines create, via the equivalent of catalytic converters, or in this instance, particle traps. Although this is correct,[94] the traps have proved less than fully successful with vehicles which cannot generate enough heat to burn off the captured pollutants. As such, their applications have often been limited to typically larger vehicles, or vehicles which can generate enough heat to operate efficiently (i.e. they do not necessarily operate well while the engine is cold).[95]

[90] Gould, R., 1989, 'The Exhausting Options of Modern Vehicles', *New Scientist*, 13 May, pp. 20–5; Joyce, C., 1980, 'Foggy Future for Diesel Cars', *New Scientist*, 9 October, p. 79; Patel, T., 1995, 'France Counts the Cost of Cheap Diesel', *New Scientist*, 22 April, p. 10.

[91] Joyce, C., 1980, 'Foggy Future for Diesel Cars', *New Scientist*, 9 October, p. 79.

[92] Cited in Lomborg, *The Sceptical Environmentalist*, p. 170.

[93] Pearce, F., 1997, 'Devil in the Diesel', *New Scientist*, 25 October, p. 4.

[94] Hamer, M., 1997, 'Fighting for Air', *New Scientist*, 19 April, pp. 14–15; Anon., 1997, 'Delightful Diesel', *New Scientist*, 7 August, p. 11; Davis, B., 1999, 'Just Add Water', *New Scientist*, 13 March, pp. 36–39; Coghlan, A., 1998, 'Clean Burn', *New Scientist*, 11 April, p. 17; Hamer, M., 1994, 'Cleaner Diesels Take to The Road', *New Scientist*, 26 November, p. 22.

[95] Thisdell, D., 1999, 'Clean Burn', *New Scientist*, 24 April, p. 12; M.

The final example in this area (I am not going to consider his other idea of having more particle pollution as a way to offset climate change) relates to his suggestion to sequestration via putting lead into the ocean, so as to increase the growth of plankton, and subsequent removal of carbon dioxide from the atmosphere. Although this idea has received theoretical attention,[96] it is important to note that there is 'considerable quantitative uncertainty' in this area due to a number of reasons.[97] In particular, it has been shown that massive amounts of seeding would be required to make relatively small reductions in carbon dioxide build-up[98] and that dumping extra iron into the oceans may disrupt ecological cycles. On the ecological side, seeding the oceans may actually encourage the bacteria that produce methane and nitrous oxide;[99] it may also disrupt the nutrients patterns near the surface of the ocean and detrimentally affect biological activity (such as with fisheries).[100] Finally, rather than resulting in an explosion of algae in the longer term, the planktonic animals that feed on the algae obtain a massive free lunch in the short term.[101] Due to such limitations,

---

Hamer, 1998, 'Clean Bill of Health for Particle Trap', *New Scientist*, 21 February, p. 6.

[96] See Fell, N., 1993, 'Can Algae Cool The Planet?', *New Scientist*, 21 August, pp. 34–7; Brown, W., 1990, 'Flipping Oceans Could Turn Up the Heat', *New Scientist*, 25 August, p. 11; Pearce, F., 2000, 'A Cool Trick', *New Scientist*, 8 April, p. 18.

[97] IPCC, 1995, *Climate Change 1994: Radiative Forcing of Climate Change*, Cambridge University Press, Cambridge, p. 17.

[98] 160,000 square kilometres of ocean, seeded with iron for a month over a 100–year period would only sequest an additional 2–11 per cent of extra carbon than would have been taken by the natural process. Jones, M., 2002, 'Don't Rely on Plankton to Save the Planet', *New Scientist*, 16 February, p. 16.

[99] Jones, N., 2001, 'A Risk Too Far', *New Scientist*, 20 October, p. 7.

[100] Gribbin, J., 1991, 'A Technological Fix That Does Not Work', *New Scientist*, 16 March, pp. 46–7.

[101] Holmes, B., 1994, 'No Quick Fix For Climate', *New Scientist*, 26 February.

sequestration in the ocean has received little attention within the climate change regime. Moreover, the Kyoto Protocol has limited emissions by sources and removals by sinks to land-use change and forestry activities, limited specifically 'to afforestation, reforestation and deforestation'.[102] As such, the Kyoto Protocol does not apply carbon accounting for 'techno-fixes' with regard to the ocean.[103]

Sustainable Development and Security

One of Lomborg's themes is that the world is becoming increasingly 'secure'.[104] When looking at the possibility of conflict over limited resources (such as water), he forthrightly discounts the event in the past (as a catalyst for conflict) or even in the future, as: 'water co-operation is highly resilient'. He suggests that war over resources such as water makes: 'little strategic or economic sense. Rather, it is to be expected that increased water will help increase the focus and attention needed to solve the remaining substantial water issues.'[105] The difficulty with this analysis is threefold. First, Lomborg assumes that conflicts are rational, and can be examined via either traditional, strategic or economic objectives. In fact, many of the current conflicts of the world are beyond these goals, in which evils like ethnic hatred, religious dogmatism, or totalitarian goals take precedence, and accordingly, the normal limitations of rational people (and international humanitarian law) do not necessarily apply. Second, Lomborg's tight focus on conflict over water fails to recognize that conflicts are typically about multiple issues, in which numerous concerns are all interwoven. Nevertheless, if viewed with Lomborg's tight prism, his point is correct (about water being the only cause of conflict). However, when considering

[102] Kyoto Protocol, Article 3 (3).
[103] IPCC, 2000, *Land Use, Land-Use Change, and Forestry*, Cambridge University Press, Cambridge, p. 5.
[104] Lomborg, pp. 84–85, 328.
[105] Ibid., p. 157.

other natural resources, it appears clear that within the last fifteen years, conflicts have involved, along with other considerations, 'honey pot' resources, as opposed to the 'shrinking pie' resources. This can be seen in Colombia (oil, gold, cocoa), Peru (cocoa), Sierra Leone (diamonds), Liberia (timber, diamonds), Sudan, (oil), Angola (oil, diamonds), Republic of Congo (oil), Democratic Republic of Congo (Zaire—copper, diamonds, gold), Cambodia (timber, gems), and Myanmar (timber, tin, gems).[106]

The above highlighted conflicts point to the third overall difficulty in Lomborg's analysis—that he fails to fully distinguish between the nature of conflict at the turn of the new century. That is, although the number of wars has clearly fallen (from 36 wars in 1989) to 27 at the end of the century (an apparent progress), 26 were internal 'civil' wars, as between opposing nation–states. It is these dynamic and often the fanatical forces that rise from it, replete with the end of the (ironic) stability of the Cold War, that has turned traditional ideas of security on their head. One of the best indicators of this is the reknowned 'minutes to midnight' clock which is set by the Directors of the Bulletin of the Atomic Scientists. The clock symbolizes how close humanity is to nuclear conflict and possible extinction. Extinction is probable if a full-scale nuclear exchange occurs. If the clock were to strike midnight, it would be the end point. This fictional clock, which was first set at seven minutes to midnight in 1945, is now back at exactly the same place as when it started—seven minutes to midnight. This is not something to consider an achievement. At the end of the Cold War, the clock was significantly wound back, but since then, it has moved forward—three times since 1991. The last move happened in April 2002, when the clock was moved from nine minutes to seven minutes.[107]

[106] Pearce, F., 2002, 'Blood, Diamonds & Oil', *New Scientist*, 29 June, pp. 36–40.

[107] *Bulletin of Atomic Scientists*, 2002, 'Seven Minutes to Midnight', April.

The clock is also moving forward because of the failure of the United States, and subsequently the international community, to accept the Comprehensive (Nuclear) Test Ban Treaty; the failure of the international community to make the (nuclear) Non-Proliferation Treaty meaningful, and tied to specific targets; the failure of the United States to maintain the Anti-Ballistic Missile Treaty, and the failure of the international community to agree to adequate verification procedures for chemical and biological weapons. In addition, there is the problem of nuclear terrorism (since 1991 there have been 18 cases of theft of weapon grade uranium or plutonium from the former Soviet Union), madmen flying jet airliners into civilian targets, and continuing border skirmishes between the known nuclear powers Pakistan and India, including an attack on India's parliament. Of late, we may also throw into the equation the highly volatile situation of a nuclear armed, provocative, and surrounded Israel.

Poverty and Prosperity in Sustainable Development

One of Lomborg's central suggestions is that a large part of the overall improvement of the state of the world is due to an ever-increasing global market. His basic point is that *on average*, the world is growing richer,[108] the number of people in absolute poverty has dropped,[109] less people are starving,[110] *most* people live longer[111] and are better educated, and there are fewer

---

[108] 'We have all become much richer than in all previous history', Lomborg, *The Sceptical Environmentalist*, pp. 70–1.

[109] Fallen from 1.3 billion to 1.2 billion at the end of the decade. Ibid., p. 72.

[110] The proportion of people starving has fallen from 35 per cent in 1970, to 18 per cent in 2000, and is expected to fall to 12 per cent in 2010. Ibid., p. 61. He correctly notes that the improvement has not been shared by sub-Saharan Africa. pp. 5, 61, 65.

[111] 'But the important thing to stress is that more than 85 per cent of all the world's inhabitants can expect to live for at least 60 years—more than

accidents[112] now as compared to 30 years ago.[113] To continue these trends, he suggests:

... it is necessary to have an open economy in order to facilitate international trade, investment and economic freedom, because this encourages the exchange of technology and administration.[114]

... if we want to leave a planet with the most possibilities for our descendants, in both the developed and developing world, it is imperative that we focus primarily on the economy and solving our problems in a global context rather than focusing ... on the environment in a regionalised context. Basically, this puts the spotlight on securing economic growth, especially in the third world, while ensuring a global economy, both tasks which the world has set itself within the framework of the World Trade Organisation ....[115]

... we need to confront our myth of the economy undercutting the environment. ... only when we get sufficiently rich can we afford the relative luxury of caring about the environment.[116]

Lomborg expects that if this path is followed, there will be a good outcome for all countries.[117] He also suggests that the countries which are following the economic medicine prescribed for them are generally doing very well in overcoming their difficulties, and responding to financial irregularities much faster and with greater strength than expected (and he cites Latin America, and in particular Brazil which shows good potential for 'stable development').[118] In following the trajectory of such development, the poorer countries will eventually eclipse their

---

twice as long as people were expected to live on average just a hundred years ago. Incredible progress', *Ibid.*, p. 53. With regard to the sub-Saharan difference, (largely put down to AIDS), see Lomborg, p. 51–2.

[112] Ibid., p. 84–5.

[113] Ibid., p. 79–113, 328.

[114] Ibid., p. 72.

[115] Ibid., p. 324.

[116] Ibid., p. 33.

[117] 'On the whole, we have no reason to expect that this progress will not continue'. Ibid., p. 330.

[118] Ibid., p. 76.

environmental problems (as did the wealthier countries before them).[119] Even if they cannot eclipse such problems and actually face difficulties in the future, they will be able to confront them via their economic prosperity. For example, with regard to possible water shortages in the future (despite strong progress in the past)[120] Lomborg suggests that shortages may be circumvented via both changing the national utilization of water, and via desalination plants. In particular:

we can have sufficient water, if we can pay for it. Once again, this underscores that poverty and not the environment is the primary limitation for solutions to our problems . . . . Desalinisation [sic] puts an upper boundary on the degree of water problems in the world.[121]

However, before even utilizing desalination plants, it may be possible to reduce water usages within an economy, via moving from high water utilization schemes, to low water uses. In principle, this means a movement away from agricultural production, as agriculture is typically water intensive (ironically, often the varieties that were heralded with the Green revolution).[122] Consequently: 'It is reasonable to expect that the most water-scarce nations will shift their availability away from agriculture and towards more valuable output in services and industry'.[123] Moreover, any shortages in food availability from this shift in production should be offset via 'extra imports by extra production in the water abundant countries, particularly the United States'.[124]

Similar conclusion, that economic prosperity (often mixed with a strong dose of technological optimism) will solve the problems facing humanity, is repeated with overall problems like climate change,[125] and interconnected problems like malaria,[126]

---

[119] Ibid., p. 176.
[120] Ibid., p. 21.
[121] Ibid., p. 153.
[122] Ibid., p. 62–3.
[123] Ibid., pp. 155, 158.
[124] Ibid.
[125] Ibid., p. 317.
[126] Ibid., p. 292.

food production,[127] and, to a degree, even with AIDS.[128] It is at this point that Lomborg has again overstated his case. In doing so he fails to understand the political economy of sustainable development. That is, although I fully concur with Lomborg's general conclusion that economic development is needed in multiple countries, I disagree with his understanding of the current economic situation, and his interpretation of the future economic prosperity for all countries, premised on the restrictive view of development that he presents.

Current Development Situation

At the United Nations Millenium Summit, world leaders agreed on a set of quantified and monitorable goals for development and poverty eradication by 2015. As Lomborg correctly points out, clear progress in moving towards these goals has been made in a number of countries and regions. However, although he does concede there are differences in the rates of success, these differences are actually more distinct problems than he concedes. Behind the record of overall progress lies a more complex picture of diverse experiences across regions in both overall indicators, and differentiation with regard to aspects of their implementation. Although the economies of some countries are clearly growing, the economies of other countries are clearly shrinking. In the period 1990–9, 10 countries saw their economies contract by more than 5 per cent per year. Only 6 countries grew at more than 5 per cent; East Asia grew much more; South Asia a little; America and the Middle East scarcely changed, and sub-Saharan Africa sunk. A large part of this growth, especially in developing countries, can be attributed to external (private sector) investment which is limited to a small clutch of countries. For example, in 1997, 10 countries took three quarters of the world's foreign

---

[127] 'It is now well recognised that failure to alleviate poverty is the main reason why undernutrition persists'. Ibid., pp. 101, 66.

[128] The problem of AIDS is 'primarily caused by political and social factors'. Ibid., p. 15.

investment; China took a quarter, while the poorest countries received little investment.[129]

In terms of the Millennium Goals, the 2001 Human Development Report, brought out by the UNDP, points out that although 66 countries are on track to reduce under-five mortality rates by two-thirds, 93 countries, with 62 per cent of the world's population, are lagging far behind or slipping. Similarly, while 50 countries are on track to achieve the safe water goal, 83 countries, with 70 per cent of the world's population, are not. More than 40 per cent of the world's population are living in countries on track to halve income poverty by 2015. Yet they are only in 11 countries (that include India and China with 38 per cent of the world's population), and 70 countries are far behind or slipping. Without China and India, only 9 countries, with 5 per cent of the world's population, are on track to halve income poverty.[130] In terms of overall development (via the comprehensive UNDP Human Development Index), Zambia had a development standard less than it had in 1975. Romania, the Russian Federation, and Zimbabwe have development standards below their 1980 levels. Development in Botswana, Bulgaria, Burundi, Congo, Latvia, and Lesotho are lower than their 1985 development levels. Belarus, Cameroon, Kenya, Lithuania, the Republic of Moldova, South Africa, Swaziland, and the Ukraine all have development levels lower than those at the start of the 1990s. Finally, Namibia and Malawi have development levels below those prevalent in these countries in the mid 1990s. Shocking results can be seen with smaller, specific indicators. For example, in 27 countries, there has been a net decline (greater than 5 per cent) in the enrolment of girls in secondary schools since the mid-1980s.[131]

As it stands, of the 4.6 billion people in the developing world, more than 850 million are illiterate, nearly a billion lack access to improved water sources, and 2.4 billion lack access to basic

---

[129] Editor, 2002, 'The Chips Are Down', *New Scientist*, 27 April, p. 31.

[130] UNDP, *Human Development Report 2001*, Oxford University Press, New York, p. 1.

[131] Ibid., pp. 10, 15.

sanitation. Nearly 325 million boys and girls are out of school, and 11 million children under the age of five years die from preventable diseases each year. Around 1.2 billion people live on less than $1 per day, and 2.8 billion on less than $2 per day. Such problems are all present in the OECD countries, with more than 130 million being classified as income poor and adult functional illiteracy rates at 15 per cent.[132]

Finally, it is important to note the often widening inequalities between the richest and the poorest, both between countries and within countries. With regard to world inequality, in 1993, the poorest 10 per cent of the world's people had only 1.6 per cent of the income of the richest 10 per cent. The richest 1 per cent of the world's people received as much income as the poorest 57 per cent. Nearly 25 per cent of the world's people receive 75 per cent of the world's income, and in the United States alone, the richest 10 per cent of the population have a combined income greater than that of the poorest 43 per cent of the world's people (around 2 billion people).[133] In total, the richest 20 per cent of the world spend 86 per cent of its wealth.[134]

In terms of inter-country inequality, a study of 77 countries with 82 per cent of the world's population shows that between the 1950s and the 1990s inequality rose in 45 countries, and fell in only 16. In the remaining 16 countries, no trend was evident.[135] In the United States, over the past two decades, following 60 years of moving towards a more equal society, the trend has been thrown into reverse. Close to 40 per cent of the national wealth is now owned by the top 1 per cent of households, very close to where it was in the 1920s, after having fallen to a far more equitable 20 per cent in the 1970s.[136]

[132] Ibid., pp. 9–10.

[133] UNDP, *Human Development Report 2001*, Oxford University Press, New York, p. 19.

[134] Editor, 2002, 'The Chips Are Down', *New Scientist*, 27 April, p. 31.

[135] Ibid., pp. 16–7.

[136] Editor, 2002, 'The Chips Are Down', p. 32.

The above situation suggests that although overall progress in certain indicators may be apparent over the last 30 years, this has not been evenly shared. In certain countries, the situation is worse, not better. Even within many of the wealthier countries, a chasm of inequality between the citizens is often developing. The question, however, is not only one of the current situation, but what will happen in the future, when global population growth moves from a current six billion, to close to ten billion, and much of this growth will be in the countries which are already suffering? That is, will the progress of humanity, overall—and in this context, in economic terms—be better or worse?

Climbing Out of Poverty?

It is essential to attempt to answer this question, as many of Lomborg's predictions rest on the assumption that as countries generally—and individuals within countries—gain more wealth, they will move away from the environmental threats that may surround them. For example, countries with water shortages will be able to solve their problems through (expensive) applications like desalination. I find this suggestion unrealistically optimistic. The countries that currently utilize desalination plants (such as Kuwait, Saudi Arabia, and Libya) are, in relative terms, quite economically (but not necessarily developmentally) wealthy due to their oil reserves. Conversely, many of the countries facing water stresses are already very poor, and by 2025 their numbers (often paralleling large population growths) will be greatly expanded. This list includes countries such as Ethiopia, Malawi, Haiti, Burkina Faso, Somalia, Rwanda, Algeria, and Burundi. The economic options for these countries are few, to say the least. To expect that they will advance to the economic positions of more resilient, but equally water-vulnerable countries such as Singapore, Egypt, Oman, or Kuwait, etc., within less than 25 years is highly unlikely. This is especially so given the current directions in the international arena, pertaining to considerations such as debt, trade, and markets.

As I have discussed these areas in detail elsewhere,[137] I shall only apply my analysis of these factors to the areas that Lomborg has advocated. First, it is necessary to say something about Lomborg's rosy view of the current economic situation. This is clearly mistaken, as in reality, often countries which are suggested as making stable progress at the time of printing a book end up being on the verge of collapse six months later. The current example of this is Latin America. In this region, following the economic meltdown of Argentina in the middle of 2002 (as of July 2002, their currency was at three quarters of the value it was at the beginning of the year) it seems quite possible that a domino effect may be seen in a number of other Latin American countries (notably Brazil, Uruguay, Paraguay, Venezuela, and Peru).[138] The key country in this pact is Brazil (the largest economy in Latin America), which despite Lomborg's earlier prognosis, is facing a debt situation (which, in terms of the public debt to GDP ratio, has risen from 49 per cent in 1999 to 55 per cent in 2002). This situation has caused leading market analysts (such as Morris Goldstein, of the Institute for International Economics) to predict a 70 per cent chance of debt default by the end of 2003.[139] The importance of debt in this area, as with multiple other developing countries cannot be underestimated (although Lomborg suggested that the debt crisis in this region had been largely dealt with). Despite repeated international attempts to solve the worst areas of the debt crisis, the overall problem has not only remained but (in total terms) has grown from 50 billion in the early 1980s to

---

[137] For a full examination of these issues, see Gillespie, A., 1999, 'Ideas of Human Rights in Antiquity', *Netherlands Quarterly of Human Rights*, vol. 17. no. 3, pp. 233–58.

[138] *The Economist*, 2002, 'A Region Prays It Will Not Slide Down Argentina's Slope', *Economist*, 29 June. Also, 'Here We Go Again', in the same issue, pp. 13; 41–2; Reuters, 2002, 'Region in Turmoil', *NZ Herald*, 28 June, p. B1.

[139] Goldstein. G., Noted in *The Economist*, 2002, 'Spreading the Risk', 29 June, p. 74.

over 1.9 trillion in 1999. As such, the debt problem is not only getting worse, it is continuing to threaten the long-term development of dozens of countries—in both economic and ecological terms.[140]

Free Trade, Free Markets, and Need

The second area that Lomborg has unwarranted faith in is with the international trading regime. Undoubtedly, international trade can be the engine for vast amounts of good results. However, when poorly structured, it may work in the opposite direction. Lomborg's idealism on free trade is clearly seen with regard to food questions. In particular, he suggests that any countries that may be detrimentally effected by climate change in terms of agricultural output, or those that may have to diversify due to water shortages, should be able to make up their shortfalls via purchasing through international trade. Given the economic predicament of many of the most likely affected countries, it is unclear exactly how they will achieve the necessary funds to achieve such goals. Despite the implicit assumptions in Lomborg's text that somehow those with more will help those with less, the reality is—and can be expected to continue as the markets evolve without any rules to the contrary—that in an international market place, the food will go in the direction of those who can pay the top price. Moreover, if a proper free trade in food is agreed, and the current subsidies in many developing countries are removed, food prices can be expected to rise.[141]

The above conclusion—that price, not need—is exactly the same with many of technological developments that Lomborg hangs his hopes on. For example, with the much heralded hopes of genetically modified (GM) crops and possible future food

[140] For a full discussion of this, see Gillespie, A, 1999, 'Ideas of Human Rights in Antiquity', *Netherlands Quarterly of Human Rights*, vol. 17, no. 3, pp. 233–258.

[141] Bongaarts, J., 2002, 'Population: Ignoring Its Impact', *Scientific American*, January, pp. 65–6.

shortages,[142] the bottom line is that GM crops will not feed the world any more than existing food supplies do, when they are tied to the vagaries of the free market and are tied to the financial directives of those with more money.[143] A very similar conclusion is to be found with some of the lesser problems that poorer countries face, such as malaria, which Lomborg suggests should be focused on, over and above other problems such as climate change. The difficulty here is not that problems such as malaria should not be concentrated on (far from it). Rather, the difficulty is that (despite a few notable public sector attempts) drug companies are reluctant to fund research on vaccines and drugs for a disease that occurs mostly in countries unable to pay for treatment. Of course, new drugs is just part of the answer. International efforts to the use of insecticide-treated bed nets and combinations of existing drugs, have strong potential in combating this disease. Such a basic control programme would cost roughly $2 billion per year. However, global spending on this (despite a World Health Organization, and United Nations General Assembly initiatives in this area) is less than $100 million per year. Set against the economic cost for Africa—an estimated $12 billion per year—this is a massive saving, but Africa, by itself, cannot afford it. And those who can are less than forthcoming.[144]

Theoretical Property and the Real World

A final example in which Lomborg places theoretical market mechanisms at the centre of his utopian vision is that of world fisheries. While admitting that some of the global fish stocks are over-harvested, he applies (in addition to fish farming)[145] the traditional argument that the solution to the classical 'tragedy

[142] Lomborg, *The Sceptical Environmentalist*, 96, pp. 342–8.

[143] Editor, 2002, 'Feed the World?', *New Scientist*, June, p. 15.

[144] Honigsbaum, M., 2002, *The Fever Trail: In Search of the Cure for Malaria*, Farrar, New York. See also, Editors, 2002, 'A Death Every 30 Seconds', *Scientific American*, 4 June.

[145] Lomborg, *The Sceptical Environmentalist*, p. 108.

of the commons' problem is to be found if 'ownership can be established over fish'.[146] I do not disagree with these conclusions, but unless they are tied into broader international legal and political debates, they fail to recognize the next stage of the debate and add little to the real problems the international community is currently facing.

The political difficulty with this theoretical analysis is that, in the real world, vast areas of the ocean are beyond the sections in which either Exclusive Economic Zones apply or strong fisheries agreements exist. As such, by the end of the twentieth century, the problem of protecting the open oceans became not so much one of 'ownership' of the resources, as one whereby some countries, or individuals within them, sort to actively violate the conservation objectives of national, regional, or international oceanic management regimes. This problem became even more complicated as the owners of vessels went 'flag shopping' (to fly the flag of countries which are not parties to fisheries agreements, and thereby avoid its conservation requirements).[147] This problem is typically recognized as Illegal, Unregulated, and Unreported (IUU) fishing, which may be consuming as much as 30 per cent of total fishery catches. In certain instances, such IUU may threaten the sustainability of parts of the resources of which international agreements are built upon. This has become a distinct problem within the Convention on the Conservation of Antarctic Marine Living Resources (CCAMLR),[148] the Commission

[146] Ibid., p. 107.

[147] FAO, 2000, *The State of the World's Fisheries and Aquaculture*, FAO, Rome, p. 13.

[148] CCAMLR Newsletter, No. 19, December 1997, p. 2. Also, Anon., 1996, 'Toothfish Pirates', *New Scientist*, 9 March p. 13; Anderson, I., 1998, 'Pirates Ahoy', *New Scientist*, 5 December, p. 8; *Illegal and Unregulated Fishing in the Convention Area*, CCAMLR XIX–2000, Commission Report, Para 5.1–5.1.6; See Associated Press, 2000, 'Antarctic Fishing Out of Control', *NZ Herald*, 6 November; See Illegal and Unregulated Fishing in the Convention Area, CCAMLR XVII–1998, Commission Report, Para 5.3. (8) Year Book of International Environmental

for the Conservation of Southern Bluefin Tuna (CCSBT),[149] and the Indian Ocean Tuna Commission (IOTC).[150] Given this reality, the problem is not of establishing theoretical 'ownership' regimes for fisheries, but mechanisms with teeth in international law, which actually protect them.

Conclusion

Scepticism is something we should all adopt. It is a wonderful philosophical tool when applied with skill. However, when applied badly, it has the potential to create more damage than good. In this area, I fear the *Sceptical Environmentalist* has done the latter. I fear this because of problems in Lomborg's advocacy with respect to data, misleading interpretations of problems, a failure to understand uncertainty, risk, and the precautionary principle, a simplistic analysis of the possibilities of technology

---

Law, 1997, pp. 267–8. (9) Year Book of International Environmental Law, 1998, pp. 265–6, 306–7. CCAMLR XVIII–1999, Commission Report, Para 5.2.

[149] See CCSBT, Report of the Third Resumed Annual Meeting, Canberra, 18–22 February, Item 11.3; CCSBT, Report of the Fourth Annual Meeting, 1996, Canberra, 8–13 September 1997, Agenda Item 8; CCSBT, Report of the Fifth Annual Meeting, Tokyo, 22–6 February 1999, Agenda Item 7, F Attachment; Action Plan Concerning Promotion of Accession to, and Cooperation with, CCSBT by Non-Member States and Entities; CCSBT, Report of the Resumed Fourth Annual Meeting, Canberra, 19–22 January, 1998, Agenda Item 2; CCSBT, Report of the Sixth Annual Meeting, Canberra, 29–30 November 1999, Agenda Item 4; CCSBT, Fifth Annual Meeting, Second Part, Tokyo, 10–13 May 1999, Agenda Item 6.

[150] See Appendix VII, Report of the Second Session of the Scientific Committee, Report of the Indian Ocean Tuna Commission, Report of the Fourth Session, Kyoto, 13–16 December 1999. IOTC/S/04/99/R[E], Paragraph 24; Appendix IX, Resolution 99/02, Calling for Actions Against Fishing Activities by Large Scale Flag of Convenience Longline Vessels, Appendix VIII, Resolution 99/01.

in these areas, and a superficial view of security. Finally, his analysis of the political economy with regard to differentiation within overall results relating to 'economic progress', problems of international debt, the structure of free trade, or the lacunas in international policy (with regard to property debates) is lacking. Unless all of these problems are addressed, there will be misleading analysis and discussions of sustainable development, and faulty foundations may be laid or further entrenched.

6

Enabling Governance for Sustainable Development

Tony Meppem, Jenny Bellamy, and Helen Ross

Introduction

Over the last quarter of a century there has been a growing and widespread concern in the global community that economic development and environmental management initiatives are not being effectively integrated. There is also widespread acknowledgement that the lack of mechanisms, arrangements, and incentives to pursue a more integrated approach for development is having substantial negative impacts on the natural resources base and consequently raises significant concerns regarding societal equity, both currently and for future generations. This is an international policy agenda that first came to prominence when these concerns were placed on the political agenda by the early work of Carson (1962), Schumacher (1974), and Meadows et al. (1972). These reports and books contributed to an altered perception regarding the trajectory of growth worldwide, from one of limitless capacity to something more bounded by resource availability, capacity of 'sink' functions of natural resources, and importantly societal equity regarding resource use.

Since then, significant change has occurred in the thinking regarding the role of governments in the management of natural resources on behalf of the society. A specific initiative that brought the resource governance agenda to the forefront of global political focus was the World Commission on Environment and Development (WCED), which led to the report *Our Common Future* (WCED 1987), commonly referred to as the Brundtland Report. This report provided guidance on how to interpret sustainable development and canvassed the potential global implications of not attending to this agenda. Sustainable development was defined as 'development that meets the needs of the present without compromising the ability of future generations to meet their own needs'. Widespread international debate ensued in the political, bureaucratic, and academic arenas, leading to the 1992 United Nations Conference on Environment and Development (UNCED), known as the Rio Earth Summit. This conference, attended by 183 world governments, articulated the implications of the sustainability agenda in more comprehensive terms and led to an internationally agreed protocol for development called Agenda 21 (UNCED 1993). The resulting mantra of 'act local think global' underpins the evolving new governance agenda for sustainable development, which is the focus of this chapter.

Australia's state and federal governments set out to address these developments in the early 1990s through an ambitious multisectoral initiative to develop a National Strategy for Ecologically Sustainable Development (Council of Australian Governments 1992). This strategy was intended to set major new directions in environmental management arrangements, including in agriculture and industry. Though not formally endorsed in its entirety, it has been influential in developing new policy directions, and the round table deliberative process it used has assisted the development of intersectoral working relationships and widespread recognition of the need to manage our enterprises and environment differently.

The interconnected nature of economic, social, and environmental issues is the perennial theme in the sustainable development discourse. The interpretation of environmental care has expanded to encompass economic vitality, social cohesion, and environmental integrity: colloquially known as the 'triple bottom line'. It has taken over a decade of policy learning, through experimentation and reflection, to advance the recognition and to gain some clarity regarding the 'framing' role of governance in practically articulating a sustainable development strategy. In broad terms, we are seeing a major movement within Australia from a construction of government and non-government bodies in a regulatory–regulated, and 'top-down, bottom-up' relationship (Carr 2002), towards collaborative roles of mutual responsibility, with government shifting more towards an enabling role. Government retains its powers, but relies more on partnerships to design, share ownership of, and achieve sustainability goals. Public participation, formerly handled as a stage in planning and environmental impact assessment, is now becoming embedded in governance processes.

The emerging governance agenda is occupied with more effective development *and implementation* of policy strategy that addresses the nested requirements of the myriad of international conventions and national and state level legislation that relate to sustainable development. This need to integrate among scales and jurisdictions, and to devolve specific decision-making closer to its context, is encouraging and enabling a greater focus on regional, catchment, or other area-based management frameworks for the development and implementation of sustainability strategies. Regions provide context for stakeholder co-operation and enable different approaches to be tailored to different geo-political, ecological, socio-economic, and cultural circumstances.

A common model, loosely adapted from collaborative planning (Gray 1989; Healey 1997), brings government and non-government stakeholders together in regional bodies (boards) that first plan, then may or may not be enabled to implement regional approaches that attempt to integrate environmental with economic and social dimensions of development. In Australia this is reflected in adaptive policy processes such as 'Integrated Catchment Management', 'Regional Forest Agreements', 'Place Management',

'Indigenous Co-Management', and 'Regional Planning'. These seek collaborations through multi-stakeholder processes (Gray 1989; Healey 1997) to provide sustainable development outcomes. Multi-stakeholder processes encompass a need to separate the roles of government as purchaser of sustainable outcomes and as provider of these negotiated outcomes (OECD 2001).

The intention of this chapter is to develop a framework for considering the governance involved in addressing complex policy issues for sustainable development. The impetus for this emergent governance agenda arises from altered societal dynamics, which include science's altering relationship with society. It is hoped that this chapter will contribute to a more comprehensive interpretation of an emerging 'sustainability science' that takes a reasoned view of the place of community and community-based sustainable development in natural resource governance (Bourke and Meppem 2000).

Studies of effective management for sustainable development have led to the identification of three interrelated research agendas for resource governance (Bellamy and Dale 2000; Dale and Bellamy 1998; Dale et al. 2001). These emphasize a focus that:

- seeks strategy for sectors to develop their own planning and management capacity;
- facilitates better collective (multisector) understanding of the social, economic, and physical processes within a particular context; and
- develops strategy to support stronger institutional arrangements that facilitate negotiation between various sector interests.

This research orientation promotes the development of strong collaborative alliances as the appropriate strategy to address complex sustainable development problems. Within a particular context, strategies will emerge through creatively facilitated participatory processes.

Governance

There is a clear shift emerging in policy for the twenty-first century that is seeking for more localized strategies to address increasingly complex and interrelated issues through community involvement and partnership development. The emergent alternative approaches to managing change are enabled through a devolution of specific decision-making powers for specific policy content to a more localized level. This enabling strategy by governments recognizes the necessity to empower context specific decision-making capacity for complex and interrelated problem issues. This trend is placing extreme challenges on conventional institutional structures and ways of communicating as new coalitions are emerging in response to these demands. This trend is characterized by an emphasis on developing partnerships, strategic alliances, and broader consultation with those who are likely to experience impacts from decisions.

The new lexicon in policy encompasses 'stakeholder participation', 'community capacity', 'agency capacity', 'negotiation', and 'partnership'. Responding to these demands requires new ways of communicating, creativity, and innovation in approaches that can foster the articulation and recognition of the mutual self-interest for sectors of collaborative endeavour. The intention being that through these avenues coalitions of interest groups are able to more clearly identify and articulate their priority concerns and engage in negotiation (or collaborative) processes with other sectors to gain agreement on collective strategy. This requires effective co-ordination or deliberative processes within interest groups and stakeholder categories: for instance, sector representatives require an imprimatur to negotiate and willingness on the part of their constituents to accept negotiated outcomes. This in turn requires consultation and broad agreement on priorities within each constituency (or community).

This international shift in policy formulation, articulation, and implementation represents a new approach that has three interdependent components, as shown in Figure 6.1.

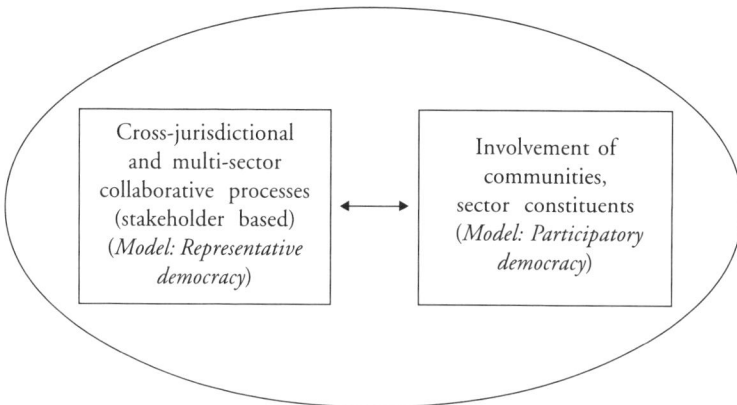

Enabling Institutional Arrangements

Figure 6.1: New Policy Environment

Contemporary governance depends on a complex set of nested relationships between different groups with a stake in the sustainable development issue (including business and industry groups, community groups, government agencies, and politicians). Its effectiveness hinges on the quality of these relationships and the extent to which different actors are able to understand the perspectives of others and have the capacity to negotiate and to undertake collective action (Bellamy and Dale 2000; Bellamy et al. 2002; Meppem 2000). This requires, for example:

- Clear agreement on the roles and responsibilities of all actors involved;
- Clarification of powers, functions, and linkages required to ensure that the cross-jurisdictional governance system is compatible with community aspirations;
- Support of inclusive, open, and collaborative forums that facilitate processes of deliberation in which networking, exchange of information, social learning, and negotiation can take place;
- Fostering of organizational/agency cultures that support community participation and are attentive to the need for change management; and

- Attention to both individual and collective/community capacity building.

We term this set of supports 'enabling institutional arrangements'.

The new role of government in these policy processes is to facilitate these new styles of interaction, to foster the emergence of creative, agreed, and context dependent policy initiatives and collaborative actions. Internationally this new approach is often called governance. This creates a dual, and sometimes confusing, role for governments. Their role is shifting from sole policy-maker to that of enabler, but meanwhile government agencies are necessarily participants in the stakeholder-based collaborative processes.

The ability for sector interests to have their issues adequately represented in this new governance agenda is highly dependent on their capacity to engage in the negotiation processes. Addressing sector (interest group or particular community) inequities in capacity to engage and negotiate is a focus for governance aimed at sustainability, hence the emerging concern for 'capacity building'.

Complex policy issues now seem to have common challenges regardless of context. These include:

- Limits of jurisdictions; complex policy issues require cross-jurisdictional collaboration for effective strategy development and implementation.
- Not a common understanding of issues; processes to collectively frame problems among stakeholders are necessary for effective implementation of strategies.
- People who should be involved are not involved; inadequate representation of stakeholder groups in problem-framing, these groups will present in-built impediments to strategy implementation.
- Tendency for planning to be generic rather than based on a particular issue; conventional planning often proposes pre-determined solution strategies based on the presumption of

understanding a problem without empowering stakeholders to articulate the nuances of a particular context.

Technology and globalization are accelerating the pace of change which is requiring governments to develop improved listening strategies to accelerate its own innovation capacity rather than relying on a more fixed solution set. The purpose of reform-through-strategies that recognize governance is to make government more responsive to society's needs. Society's desire for engagement in public policy is changing as it is becoming more diverse, complex, and multifaceted. These responses are raising the importance of considering underlying values that shape strategies for the governance of public resources. This complexity encourages a learning approach (Meppem and Gill 1998). This arises from a more widespread recognition that we (society generally) do not really know the desired outcome other than 'better than now' which requires people to collectively address and determine what is better, and the incremental steps required to set this in train.

There is concurrent reconstruction of the notion of 'management'. In environmental management, there is increasing recognition that most stakeholders 'manage' the environment, either on-ground or through influencing processes such as policies and financial incentive programmes. Landholders manage environments by the nature of their using them, stewardship groups contribute by activities such as restoration work, and governments—conventionally regarded as 'managers'—influence the activities of those with on-ground opportunities to change landscapes through regulation, policies, and programmes.

Government has become just one actor, among many, seeking to represent and provide services for sustainable development. The state does not have and cannot be expected to have all the ideas, but it does have a specific role in facilitating their emergence. The role of government becomes that of an enabler, to support the capacity for society to learn rather than seeking to direct outcomes through programmes. This entails negotiation among different interests, with government taking a strategic role including bringing parties together. Government therefore needs to be more strategic in positioning its engagement. More strategic reform requires developing a collective vision among stakeholders, which leads to the building of a constituency. Negotiated planning to achieve outcomes and communicating among stakeholders is an ongoing orientation for these strategic engagement processes. Effective dialogue processes are the currency for this altered organizational approach.

Given the nature of the tasks described for governance, outcomes are usually described more in terms of progress in a process, than through specifically quantifiable economic or physical outcomes that can be identified in advance of a collaborative approach. Improved process will, however, translate, via often complex pathways (or multiplier effects) to improved economic, socio-cultural, and environmental outcomes. The role of governance can be interpreted as promoting the working among different groups to assist in determining what the problems are and who might do what in a process of improvement. Rather than interpreting this activity as efficiency in the use of scarce resources—which is really about doing the same with less—it is more about enhancing the effectiveness of activities actually undertaken. The implementation of such strategies offers significant challenges, as this usually requires a shift in the working relations among those groups and organizations involved in activities (Meppem and Bourke 1999). It certainly requires appropriate evaluation assessment mechanisms; particularly evaluative mechanisms that can focus on the processes and their relationship to the consequences of those processes as co-dependents (Bellamy et al. 2001; Bellamy et al. 2003).

To assist in interpreting the implications of governance for sustainable development a simple diagrammatic framework will be employed. In Figure 6.2 the classic Ecosystem Model is portrayed.

Figure 6.2 emphasizes the importance of interpreting the dynamics and interrelations among various components of the

Climate/atmosphere

Biodiversity

Soil/water
dynamic

Figure 6.2: Ecosystem Model Representation

natural environment for understanding a healthy ecosystem. Put simply, the relations among climate/atmosphere, biodiversity, and the soil/water dynamic capture the complex functioning of the ecosystem.

We propose that this description of the natural environment may also be used as a metaphor to describe management or governance for sustainable development. In Figure 6.3 the ecosystem model is now labelled as the Governance Model.

In this model, the three essential components of a healthy governance system are:

(i)   Enabling institutional arrangements: The structures, legislative framework, and processes that create the 'climate or atmosphere' to encourage adaptation, empowerment, and partnership development;

(ii) Diverse sector participation and co-dependence: In the same way that less visible organisms contribute to biodiversity, a healthy governance system requires diversity of sector participation and community engagement;

(iii) Individual sectors must have processes for maintaining strong relations and feedback arrangements with their

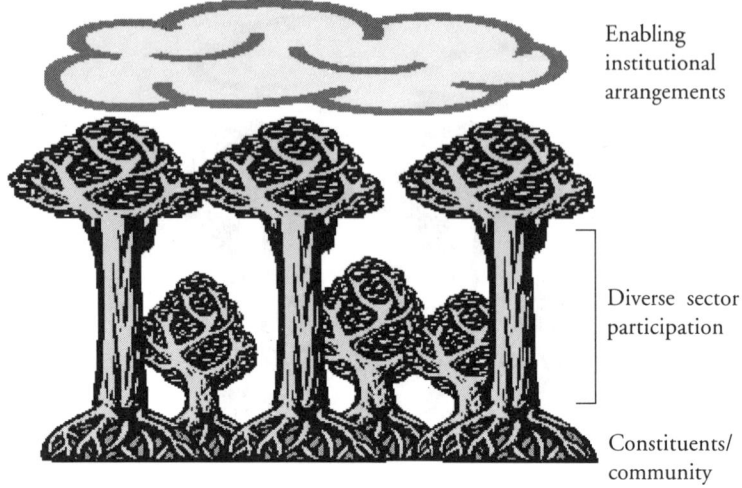

Enabling
institutional
arrangements

Diverse sector
participation

Constituents/
community

Figure 6.3: Governance Model

constituents or community. Just as plants depend on their
roots, soil, and moisture, sectoral organizations that claim
representation in sustainable development must gain their
mandate or 'nutrients' from their constituents and
communities in an ongoing way to enable the capacity to
negotiate outcomes. This requires an adaptive learning
policy strategy to enable feedback to occur so that sectors are
able to adapt and evolve in response to 'environmental
stimuli'.

With the use of this simple imagery, the components of
healthy governance can be viewed as a system. Its use can now
assist us in developing and interpreting more appropriate policy
for sustainable development. For example, in Figure 6.4 the
structural or 'jigsaw' approach to environmental management is
portrayed.

Here those sectors that are prominent or most powerful, or
whose 'tops stick out of the canopy', are most readily selected
to be on a management committee or board for environmental
decision-making, while less evident organizations, individuals, or

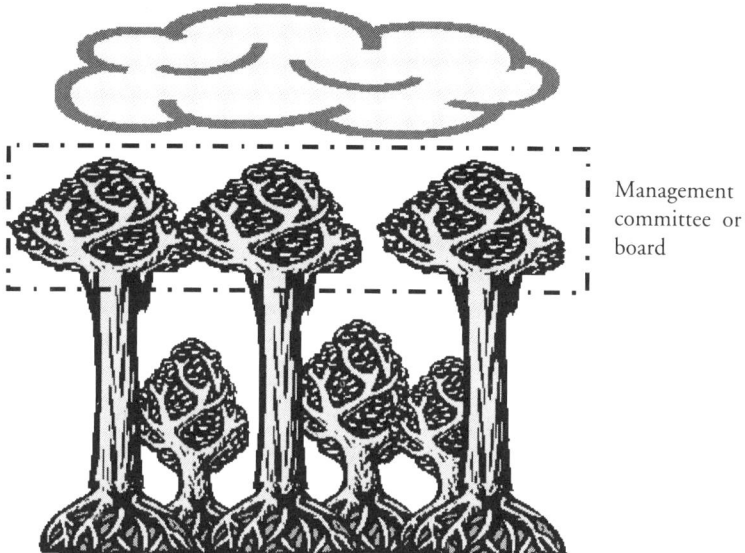

Management committee or board

Figure 6.4: Structural Approach for Environmental Management

constituencies may be omitted. This often leads to significant problems. There may be resistance from those excluded, or the decisions may be impoverished by lack of diversity in the inputs. There may be further difficulties in moving from planning to implementation, because of the lack of attention to the role of 'enabling arrangements' to encourage negotiation among diverse sectors, and also a lack of attention to the dynamics of internal communication for sectoral policy development. Strategy development is necessary within each sector to build and maintain a mandate for wider negotiatory processes to be effective. In a healthy governance system, policy negotiations are concurrently occurring both across and within sectors, with these activities being fostered by enabling institutional arrangements.

Enabling institutional arrangements typically consist of legislative and policy frameworks that assist societal deliberations toward valued outcomes; healthy organizations with excellent constituent relationships in two-way communication flows (vertical integration); healthy communication with mutual respect among sectors (horizontal relationships); equitable deliberative processes leading to widely accepted outcomes; and socially robust knowledge (Gibbons 1999).

The environmental policy domain has shifted from using good scientific information as a beacon on the horizon providing long-term direction for strategy. The concept of 'socially robust knowledge' makes room for assertions of 'sense of place', amenity, environmental integrity, biodiversity, indigenous spirituality and cultural responsibility, and intergenerational/intersectoral equity. It reflects an altered orientation for knowledge in complex contexts from what is 'known' to the relations between forms of system knowledge and what society collectively wants to 'do' to influence outcomes based on this 'interpretative' knowledge. The 'what' to do in complex contexts acts in a symbiotic relationship with 'how' this 'what' is determined. This has significant implications regarding representation in sustainable development decision-making since process and outcomes are interdependent.

Governance of Water Resources

Let us briefly illustrate these arguments through a complex case, that of governance of water resources. Allocation of water resources, for instance in a river, brings into play:

- Substantive arrangements, such as regional planning schemes, water allocation plans, and other legislative arrangements including property rights that form the legal and administrative structures for water management.
- Stakeholders/sector groups, that is the interests that affect, or are affected by, the distributions of water (including organizations and constituencies). Various kinds of knowledge inform sector planning and policy development, for example: technical knowledge in the form of catchment and ecological flow modelling and integrated ecological information systems; practical knowledge from the experiences of resource users;

and diverse community including indigenous knowledge from those who experience flow-on impacts from water management practices.

- Deliberative processes, such as stakeholder-based planning, or negotiations with a focus on communicative practices to develop and maintain sector constituency for effective negotiation in water management.

It is through the interactions of diverse stakeholders/sectors in the context of substantive arrangements and deliberative processes that socially robust knowledge emerges for decision-making to move to sustainable outcomes in the water sector. Our simplified representation using the 'Governance Model' is a means to conceptualize the dynamic interrelationships and co-dependence of an effective and adaptive water management governance system. Being better able to interpret the relations among the three core elements of a healthy governance system enhances the capacity to manage the system in a more effective and holistic way. A recent study in catchment management in Australia illustrates the complexity and co-dependent nature of these elements in practice (Bellamy et al. 2002).

Enabling
institutional
arrangements
(*Substantive*)

Diverse sector
participation

Constituents
(*Deliberative*)

Figure 6.5: Effective and Adaptive Water Governance System

Concluding Remarks

The implementation of resource governance arrangements is inherently context-sensitive. A wide range of environmental, economic, social, policy/institutional, and technological factors will influence its implementation and impact. Natural resource management and economic development activity is embedded in societal associations through established social networks and interactions, social values, institutional frameworks, historical problems, past experiences, and established ways of doing things, and it will continue to be shaped by them in the future. Resource governance arrangements need to be able to adapt and to suit local contexts. Evolution in institutional arrangements is therefore vital to ongoing improvement in resource governance. Building the capacity of society to engage in participatory process is a fundamental role for government in planning for sustainable development. This will involve supporting and strengthening existing networks, shifting 'communicative lines of sight' which will influence the operation of institutional arrangements, and facilitate the use of scientific, indigenous, and community information into decision-making frameworks, rather than constantly seeking to create a new set of structural arrangements. This defines a role for government that is away from regulator or provider of technical expertise and toward enabler, facilitator, and partnership negotiator.

There is a need to produce tangible sustainable development outcomes through our efforts in water sector management. This position asserts three interrelated dimensions for reassessing the sustainable development paradigm. First, the ability for sectors to have their own agreed policy and planning process is essential in developing the capacity to negotiate for sustainable development. This is necessary for defining boundaries, modes of operation, and priority issues. Without this within–sector capacity to engage in negotiation, it is not possible for more broadly representative governance arrangements to effectively move from planning to implementation. Second, governance arrangements should provision a framework that draws in complexity and allows for intersector collaborative assessment of how the various aspects of change relate to each other in a management context. The emphasis in this context is on the relationships among the components impacted on by change. The intention of the evolving framework agreements of governance is to provide greater certainty regarding the scope and intent of planning and implementation processes. Third, arrangements and processes to effectively mediate communication are crucial for capacity building to align commitment and engage collaborative endeavour for mobilizing sustainable development. All three of these interrelated dimensions are essential elements of a healthy governance system for natural resource management and sustainable development.

Glossary

Resource Governance can be interpreted as the interplay among institutions, legislation, information, communications, power, perceptions, and interests that are currently shaping our responses to environmental issues and consequent decision making.

Landscapes include people. Landscapes evolve through the interactions between human influences, the built environment, and 'natural' landforms, biota, and climate. Landscapes have public amenity values beyond their utilitarian value. Landscapes are inherently diverse. Landscapes are socially constructed.

Sector is a term used along with community to denote a specific category of special interest group, for example, government sector, indigenous sector, business sector, conservation sector.

Region: A regional approach can be understood as addressing issues that are multi-aspect or multi-disciplinary, encompass cross-jurisdictional implications, and have diverse sector impacts. This categorization recognizes the different scales in regional approaches and places a greater emphasis on the relations between actors and their organizations as the focus for regional approaches.

Organizational culture is the embodiment of all the recurring practices in the way people work, interact, co-operate, make decisions, and resolve problems. It is notoriously more difficult to change these working methods than to alter structures, which is generally more symbolic than innovation in practice.

Institutions: Desired changes in on-ground management are influenced by a complex mix of policies, land tenure, property rights, incentives, market forces, planning, administration, regulation, and so on. The shorthand term we use to refer to these elements of the operating context for natural resource management (NRM) is institutions. The institutional dimension of NRM thus refers to the way in which society organizes itself through relatively formal structures, policies, rules, and procedures.

Sustainability: It has been widely accepted for some time that the concept of sustainability cannot be reduced to a simple, absolute statement or definition—it is a journey rather than a destination. Outcomes will therefore be socially constructed, weighing up ecological integrity with social equity and economic efficiency, riven with tension, compromise, and non-linear trade-offs. The real challenge is not to define sustainability, but to develop the processes, the forums, the modes of inquiry and learning, to better inform and support debate, planning, decisions, and actions. The values people or societies ascribe to landscapes, whatever those values have a major influence on the way they interact with and manage them. Policy for sustainability needs to:

- Contribute to an enabling environment for diverse engagement in complex issues;
- Reward governance and encourage a strategic consensus on policy issues;
- Recognize nested clusters of stakeholders and interest groups who develop joint positions and to then enter into dialogue with the other main players;
- Seek to solve many contentious issues in the pre-negotiation phase; and

- Address place based outcomes rather than agency specific outcomes.

Stakeholders: These are parties who affect, or are affected by, any issue or decision, who have a 'stake' in the outcome. This is an analytical category, and can include abstracts such as 'future generations'. A stakeholder category, such as 'conservation groups' or 'local communities' may or may not correlate to identifiable organizations. This contrasts with 'actors', actual individuals, and organizations—the set of actors in an environmental issue may not give complete coverage of stakeholder categories (some may be left out of the process or be unable to organize themselves sufficiently), and there may well be several actors within a single stakeholder category.

References

Arthur, W. S., 2001, '*Indigenous Autonomy in Australia: Some Concepts Issues and Examples*, Centre for Aboriginal Economic Policy Research, Discussion Paper No. 220/2001, The Australian National University, Canberra.

Bellamy, J., and A. Dale, 2000, *Evaluation of the Central Highlands Regional Resource Use Planning Project: A Synthesis of Findings*, Final Report to Land and Water Australia, R&D Project CTC13. CSIRO Sustainable Ecosystems, Brisbane, November (http://chrrupp.tag.csiro.au).

Bellamy, J., H. Ross, S. Ewing, and T. Meppem, 2002, *Integrated Catchment Management: Learning from the Australian Experience for the Murray-Darling Basin*, Final Report, A Report for the Murray-Darling Basin Commission. CSIRO Sustainable Ecosystems, Brisbane, January (http://www.mdbc.gov.au/naturalresources/planning/icm/icm_aus_x_overview.html).

Bellamy, J., D. H. Walker, G. T. McDonald and G. J. Syme, 2001, 'A Systems Approach to the Evaluation of Natural Resource Management Initiatives', *Journal of Environmental Management*, vol. 63, no. 4, pp. 407–23.

Bellamy, J., T. Meppem, R. Gorddard, and S. Dawson, 2003, 'The changing face of regional governance for economic development: Implications for local government', *Sustaining Regions*, vol. 2, no. 3., pp. 7–17.

Bourke, S. and T. Meppem, 2000, 'Environmental Narratives and Fictions of Consent in Environmental Discourse', *Local Environment*, vol. 5, no. 3, pp. 75–82.

Carr, A., 2002, *Grass Roots and Green Tape: Principles and Practices of Environmental Stewardship*, The Federation Press, New South Wales.

Carson, R., 1962, *Silent Spring*, Riverside, Cambridge, Mass.

Council of Australian Governments, 1992, *National Strategy for Ecologically Sustainable Development*, Australian Government Publishing Service, Canberra.

Dale, A. and J. Bellamy, (eds), 1998, 'Regional Resource Use Planning in Rangelands: An Australian Review', Occasional Paper No. 06/98, *Land and Water Resources Research and Development Corporation*, Canberra.

Dale, A., J. Bellamy, and A. Leitch, 2001, *Central Highlands Regional Resource Use Planning Project: A Planning and Learning Experience (CHRRUPP)*. CSIRO Sustainable Ecosystems, Land and Water Australia, Canberra. (http://www.lwrrdc.gov.au/downloads/PK010095.pdf).

Ewert, A. W., 1996, *Natural Resource Management: The Human Dimension*, Westview Press, Boulder, Colorado.

Gibbons, M., 1999, 'Science's New Social Contract with Society', *Nature*, vol. 402, Supplement 2, no. 1, December, pp. 4–12.

Gray, B., 1989, *Collaborating: Finding Common Ground for Multiparty Problems*, Jossey-Bass, San Francisco.

Healey, P., 1997, *Collaborative Planning: Shaping Places in Fragmented Societies*, Macmillan, Basingstoke and London.

Lubchenco, J., 1997, 'Entering the Century of the Environment: A New Social Contract for Science', *Science*, vol. 279, no. 3, pp. 491–7.

McDougall, S., 2001, 'Report to Seaforum: Native Title Implications of Seaforum Activities', December, (www.seaforum.org).

Meppem, T., 2000, 'The Discursive Community: Evolving Institutional Structures for Planning Sustainability', *Ecological Economics*, vol. 34, no. 1, pp 223–45.

Meppem, T. and S. Bourke, 1999, 'Different ways of knowing: A communicative turn toward sustainability', *Ecological Economics*, vol. 30, no. 2, pp. 129–38.

Meppem, T. and R. Gill, 1998, 'Planning for sustainability as a learning concept, *Ecological Economics*, vol. 26, no. 2., pp. 190–207.

OECD, 2001, *Local Partnerships for Better Governance: Territorial Economy*, Organisation for Economic Cooperation and Development, Paris.

Sanders, W., 2002, 'Towards an Indigenous Order of Australian Government: Rethinking Self-determination as Indigenous Affairs Policy', Centre for

Aboriginal Economic Policy Research, Discussion Paper No. 230/2002, Australian National University, Canberra.

Schumacher, E. F., 1974, *Small is Beautiful: A Study of Economics as if People Mattered,* Abacus, London.

Tuan, Yi-Fu., 1974, 'Space and Place: Humanistic Perspective', *Progress in Geography,* vol. 6, no. 1, pp. 211–52.

UNCED, 1993, *Agenda 21, United Nations Conference on Environment and Development,* Rio de Janeiro, 1992.

WCED, 1987, *Our Common Future: Report of the World Commission on Environment and Development* (Brundtland Report), Oxford University Press, Melbourne.

Webber, J., 2000, 'Beyond Regret: Mabo's Implications for Australian Constitutionalism', in D. Ivison, P. Patton, and W. Sanders (eds), *Political Theory and the Rights of Indigenous Peoples,* Cambridge University Press, Cambridge, U.K.

7

River Basin Management as a Way to Sustainable Development in Latin America

Axel Dourojeanni

Introduction

One of the issues currently raising concern in water resources management is what is known as 'governability', a concept that is related to the ability to make decisions in a participatory manner and to implement them. In order to make appropriate decisions, with a view to achieving integrated water resources management goals, it is necessary to reconcile the interests and dynamics of the local population with the conditions and particular dynamics of the environment in which they live, especially with regard to river basins and the hydrological cycle. This means that decisions taken should incorporate understanding of human behaviour with the characteristics of the environment in which they are implemented. This necessary linkage of the knowledge provided by the so-called 'soft sciences' (such as sociology, anthropology, law, economics, and politics) with the so-called 'hard sciences' (physics, chemistry, biology, ecology, and engineering) rarely occurs in practice.

The lack of co-ordination mechanisms for combining the contributions of both groups of disciplines and sciences is one of the reasons for ungovernability in integrated water resources management. Decisions are usually made in a simplistic and piecemeal fashion, using paradigms established beforehand, and in most cases the decision-makers are unaware of the natural environment in which such decisions are implemented. Though proposals normally take social and physical factors into consideration, they fail to integrate these perspectives. In the social arena, for example, many initiatives advance the view that, in order to make better informed decisions, there is a need to construct a water culture, to build a level of awareness, and to formulate a policy on the importance of the resource, so that the public makes 'rational' decisions concerning water use. In light of these ideas, the assumption seems to be that there exists no culture, no policy, nor any degree of awareness of water resources management at the time the proposal is formulated, and that one needs only to establish a policy, a culture, and a level of awareness for progress to be made.

Culture is the way in which people express themselves in their social and spiritual relationships and with their environment. Culture is the way in which humans relate to the world and make decisions in order to improve their quality of life. This relationship requires a long learning and adaptation process linked to the territories people occupy or use at a distance. Culture, therefore, is linked to the accumulation of knowledge. At this point, however, it is important to shatter the first myth: traditionally, 'culture' associated with the accumulation of knowledge is viewed as an asset for human development. Acquired knowledge, including that represented by cultural characteristics, may, however, serve to impede adaptation to new situations. Societies and individuals are reluctant to change their attitudes; hence, with globalization intensifying changes and cultural shocks, we are witnessing increasingly contentious situations between the new inhabitants, long-standing inhabitants adopting new customs and the environment they live in.

As good intentions would have it, decisions must be 'rational', though no explanation is given as to what that actually means. A rational being may be defined as one who makes balanced decisions consistent both with knowledge of the environment in which decisions are to be applied and with his or her ability to carry them out (Röling 2000). Therefore, rational thinking associated with a particular culture requires knowledge and time for adaptation. In the past 50 years, the cultural shocks engendered by migration and exchanges between inhabitants of territories, including within the same country, have become even more intense. These migrations—of people as well as knowledge and technologies—alter the rational basis of decisions, since the processes of transculturization and globalization outstrip the societies' ability to adapt to the new environment in which they intervene. In addition, their impact is being increasingly widely felt due to the use of technologies of greater power. Society's inability to avoid conflicts over water use and to deal with the effects occasioned by the unpredictable nature of the resource is one result of this situation.

If we accept that action should be taken to gear cultural attitudes to the new environment that they find themselves in, three questions arise: first, what 'kind of culture' is it hoped to develop when programmes are launched to that end?; second, how much is known about the environment to be modified?; and third, how should both aspects be linked? One of the problems observed is that the majority of the so-called processes of 'culturization' and 'awareness-raising' as regards water are associated with piecemeal approaches or ones established earlier in other places and under different conditions. Thus, for example, attempts are made to educate the public about economic matters, so that they make decisions on water resources management based exclusively on market prices and so that they respond to 'economic instruments'. This approach is certainly useful, though within certain contexts. It is not necessarily more 'educated' to think solely in terms of economic values while ignoring social and environmental aspects. Reducing human reasoning to economic reasoning is not a particularly suitable culturization process for relating to the world.

Niels Röling, a well-known Dutch sociologist, believes that the instrumental and economic solutions that have brought human beings into conflict with the environment and led them to plunder will not by any means be the only solution to overcome this dilemma. In fact, says Röling (2000), it is these solutions which created the problems. Technology and economics may help build a sustainable society only if they are put to use within a framework of thinking and collective action that is superior to limited instrumental and economic reasoning. Such collective thinking does not appear to be a factor in most of the decisions taken today despite statements to the contrary in official speeches, laws, and constitutions. Every day, the public and private sectors, and society at large, express opinions in favour of the environment and social equity (for example, promoting 'sustainable development'), but the decisions they make belie their opinions. All governments issue statements expressing 'the need to attain sustainable development through participatory, democratic and multidisciplinary decisions incorporating gender and ethnic perspectives, among others,' but decisions are usually made based on the same old criteria.

In Röling's view, in order to break free of this impasse and act 'rationally' in regard to water resources management, the decision-maker must be able to link the soft sciences and the hard sciences. It is easy to see that Röling's assertion is correct, but simply aspiring to adopt a more socially and culturally focused approach, one that is interdisciplinary, participatory, and holistic, is not the same thing as actually achieving it; in a similar vein, it is not sufficient to consider the thinking of society without weighing it against the limits imposed by nature and our knowledge as to how it behaves. Here another paradigm arises. People often believe and assert that being well-informed about the environmental impact of decisions on land and natural resources development and use provides a sufficient basis on which to make correct decisions. However, human beings are slow to react to knowledge about the impact of their actions. For instance, many natural disasters are actual facts due to the population's failure to pay heed to the threat of danger by settling in flood-prone areas. The paradigm that information always leads to sound decision-making is also therefore subject to qualification. It is only valid in a context characterized by a combination of conviction and the potential to implement recommendations and guidelines on prevention.

In summary, therefore, the design of functioning integrated water resources management systems and the attainment of governability in order to secure this management requires, first of all, breaking away from paradigms, myths, and beliefs that, though valid in theory or in isolated cases, lose their validity in real world situations that are much more complex. And second, there needs to be acceptance that, in order to make 'rational' decisions, interdisciplinary working methods must be used that foster appreciation and respect for the contribution made by all sciences and disciplines, both hard and soft. Such methods exist and are available for organizing interdisciplinary activities and making participatory, transparent decisions. It is regrettable to note that for now many of the failed attempts at achieving integrated water management goals are due to the use of approaches that are piecemeal, applied out of context, and even naive.

Integrated Water Resources and River Basin Management in Latin America

Sustainable Development and River Basins

The sustainability of development remains an academic concept unless it is linked to clear objectives that must be attained in given territories and to the management processes needed to achieve this. Management of the natural resources located within the area of a river basin is a valuable option for guiding and co-ordinating processes of management for development in the light of environmental variables. In order to turn environmental policies into concrete actions it is necessary to have suitable management bodies, which are normally very complex. The establishment of such bodies means generating a mixed public and private system which should not only be financially independent, socially oriented, and sensitive to environmental aspects, but must also act in a democratic and participative manner. In the past, the idea of establishing bodies to guide the management of the natural resources of a river basin (especially water, of course) has aroused the interest of the countries of Latin America and the Caribbean, with varying results. This interest has now become an urgent necessity, in view of the greater competition for multiple water use and the need to control water pollution and manage the environment correctly.

Sustainable development does not refer to a tangible and quantifiable goal to be achieved in a given period of time, but rather to the possibility of maintaining a balance between factors that explains a certain level of development among human beings, a level that is always transitory, evolving, and, at least in theory, should always lead to an improvement in the quality of human life. Sustainable development is thus the result of a set of decisions and processes which have to be carried out by generations of human beings, under ever-changing conditions and usually insufficient information, subject to uncertainties, and with goals which are not shared by a population that is showing a growing trend to individualism.

One of the biggest concerns at present, at least to judge from policy statements, is to find viable development options based on equitable and lasting economic growth. The latter consideration has gained in importance in recent years because of the realization that many alleged advances, especially in terms of changing production patterns, have been outweighed by the damage they cause to the environment. The greater awareness and understanding that now exists of mankind's interaction with the environment, and the vulnerability of forms of development which do not take this into account, have been made more explicit by the addition to the word 'development' of the qualifying adjective 'sustainable'. Since sustainability should be implicit in the very concept of development, this adjective should be only a transitory addition that will be needed only until the vital importance that development should be of a lasting nature is definitively incorporated in the concept.

On the other hand, the sustainability of development remains only an academic idea or abstract aspiration unless the concept is linked both with clear objectives that must be attained within a given area that contains the natural elements and resources needed for the subsistence of the human race and with the management processes needed to achieve those objectives. Thus, political intentions must be transformed into concrete policies for implementation, and it is here that the greatest challenges arise.

In the Latin American and Caribbean region, there has been widespread reference to environmental problems, theories have been put forward on environmental issues, laws have been enacted, and even some ministries of the environment have been set up. What has not been done, however, is the laying of the necessary bases for the management of each of the natural resources—water, soil, forests, fauna, minerals, and energy—or of certain key natural areas such as coastal zones, river basins, and deserts.

This means that very broad goals have been set without deciding on the necessary steps for reaching them. Territorial organization for the management of each resource and later of the environment in general; organization and training of the population; research on ecosystems; the establishment of systems of management for given areas; the strengthening of public institutions (especially local governments) to provide support for environmental management; awareness and heightening of the economic value of natural resources; the keeping of natural heritage accounts; and the preparation of operating manuals and rules are essential aspects for making real progress in the management of natural resources and the environment in general.

The management of natural resources in the context of the dynamic evolution of a river basin, more generally known as river basin management, is one of the possible options for organizing the participation of users of natural resources within the process of environmental management. A river basin is uniquely fitted to serve as the basis for the co-ordination of the actions of all those involved in the use of a shared resource—water—and for the evaluation of the effects of environmental management measures on that resource. Water quality largely reflects the environmental management capacity within the basin in question.

A first step towards river basin management is to limit action to the management of the water resources existing within the area of the basin. Water management is a complex process designed to control the cycle of a natural resource whose availability is erratic and irregular over time and space. Furthermore, water is vulnerable to the treatment it receives, since it can easily be polluted, thus affecting all its actual or potential subsequent uses. The aim of this process is to solve conflicts among multiple users who, whether they like it or not, depend on a shared resource. Consequently, even though they may have water use permits or rights, they nevertheless affect and depend on each other. The supply usually comes from a common system, to which surpluses and wastewater are returned. Surface, ground, and atmospheric water resources, together with the areas where water is diverted and returned, thus form a single unit.

The actions taken have enormous repercussions on human health, the environment, and production, so that they must be approached in an outstandingly technical manner. The high cost of the works involved, together with the long lead times of water projects, make it all the more necessary that management should be in the hands of experts whose tenure does not depend on political changes.

Finally, the water management process requires that many different agents should act in a co-ordinated manner in spite of their differences of approach and the fact that some of them are not aware of the effects of their decisions on the hydrological cycle. This is why it is so important to have stable co-ordination mechanisms and, at the very least, a permanent river basin centre or authority.

Between Ideas and Facts in Integrated Water Resources and River in Latin America

A recurrent theme at recent international forums has been the so-called 'global water crisis' which stems apparently more from a mismanagement, or lack of it, of water resources and the spectre of an increasing water shortage. Discussions focus then on the lack of appropriate mechanisms for resolving conflicts among water users with respect to quantity, quality, and time.

Since average annual rainfall in Latin America and the Caribbean is estimated at about 1,500 millimetres, which is over 50 per cent above the world average, reference to a water shortage in that region in the absolute physical sense is not very appropriate, although one must recognize the fact that the natural distribution is highly uneven. There is no denying, however, that water management systems are often poorly organized if not non-existent. This results in a lack of information on water balances, an almost total lack of control of water quality and, scant preparation for natural disasters such as droughts and floods. In addition, the region still lags behind in drinking water supply and above all, in the development of adequate sewerage and drainage systems. Furthermore, deficiencies persist in the water management systems for agriculture, both in irrigation and drainage, one of the outcomes of a long tradition of State paternalism.

A major obstacle to the improvement of water resource management is the institutional legacy of systems, which traditionally were centralized, and in many cases assigned to a user sector such as agriculture or energy. Even today, there are draft bills which suggest that national water, or water resource councils should be run either by the agricultural sector or the energy sector. In addition, very often, such councils claim that they should be composed only of public officials, without water users or civil society, in particular the municipalities, having any say in the decision-making process.

Another important obstacle is the legacy of a 'subsidy' culture for water projects (main hydraulic structures) and rates. This is especially true for the irrigation projects. In fact, almost none of the costs of the hydraulic structures to improve irrigation schemes are being recovered. In many countries, there still exists resistance to the establishment of real prices for water use. In part this is due to the fact that the money collected goes to the central budget. Stakeholders never know if the money collected is being wisely used for water management purposes. The lack of transparency in the financial management is a real obstacle.

For these reasons, much of the work in Latin America should be geared towards advising governments to formulate their water policies, in an attempt to achieve a balance between the advantages of markets and private participation, particularly in drinking water supply and sanitation, hydroelectricity, irrigation and drainage, and the need for government regulation to achieve social and environmental goals. Another part of the work consists in assessing the market's real capacity to act as a mechanism for ensuring the efficient use and transfer of water rights.

Proper policy-planning requires information on existing water resources, the existence of a land register and public register of water users, effective control of water quality, and a system of participatory management at the river basin level. To achieve such results, it is necessary to improve the institution in charge of water resources management, especially at the river basin and at the agricultural levels since other users such as energy and water supply are better organized.

Today, in Latin America and the Caribbean, the water issue is immersed in a series of plans related to integrated environmental management goals, an aim which assumes that the capacity to manage multiple water use will be achieved as a by-product. For the sake of this idea, in more than one case the existing capacity for water management has been reduced in the process of adapting it to 'integrated environmental management'.

There is currently a wide range of situations, many of them going backwards in the region with regard to proposals for legislation, standards, technical specifications, organizations, capacities, research, education, and effective application of processes for multiple water use management, even within a single country. Due to this, progress is slowly being made in the consolidation of some bodies for integrated water management, both at the national level and at the level of states, provinces, and river basins.

Brazil and Mexico are the only two countries with specific mandate in the water law to create river basin authorities. Unfortunately, it has to be recognized that, in more than one case, at the national level, and at the level of river basins, government capabilities have been reduced. The current reduced governing capacity for multiple water use management is obviously not exclusively due to its being diluted by incorporation into the broader environmental issue. There are deeper causes that have existed for decades: some originate with the public and private officials responsible for water management and use; and others are external, and stem from the socio-economic situation of the countries or the river basins where the water resources are managed.

In contrast, and paradoxically, it is interesting that although almost all countries of the region agree that some organizations are needed for the management of water at river basin levels, progress has been very slow. It is certainly not a simple task, nor does the relevant legislation often exist to create such organizations with due legislative and financial support. It is thus important to create or enhance the region's capacity to support these initiatives.

The creation of a water management organization at the river basin level does not guarantee its continued existence as it requires continuous support for its consolidation in the form of technical assistance and financial resources for at least a decade. Many of the laws establishing these bodies do not provide for clearly defined roles, or the assignment of legal status, stable sources of income, personnel training, and in general the methods, criteria, standards, and operational procedures are not prepared beforehand in order to formulate plans and standards with due legality.

It is therefore suggested, as part of the necessary task of improving multiple water use management, that funding be provided for an appropriate number of researchers to systematize and standardize the experiences available. This would be possible if one or more research or logistics centres[1] were established in Latin America for multiple water use management and integrated river basin management. These centres could be set up with the support of interested organizations and could be attached to a university or some existing regional or international organization, in order to serve as an information centre both for regional water resources networks and for educational centres to support manager training for river basin organizations and multiple water use management.

Although sizeable networks do exist now for issues relating to integrated water management, there are still very few regional studies available in this area and there is even less access to criteria, standards, procedures, and working methods at the river basin level. The above centre should help move on from the present situation of dispersed information, confused ideas, lack of follow-up on the progress made, and the generally unstable procedures for training, consolidation, and functioning of water management bodies at the river basin level, and organizations for multiple water use management in some countries of the region.

Water resources management has come to figure as a major item on the agenda at national and international meetings, after a long period during which the issue went relatively unnoticed among other environmental topics. Recent international meetings, however, have neither touched on new aspects nor, apparently, made any major strides in improving water management policies, particularly not in the Latin American and Caribbean countries. The concerns of national, regional, provincial, and local governments have proved to be a great deal more elementary than the lofty declarations of principles that usually arise from these events.

Case studies and experiences that already exist indicate the need for preparing a preliminary draft water legislation, evaluating

---

[1] One proposal is to establish one centre for the Andean region (in Cuenca, Ecuador), one for Central America, one for Mexico, one for a Caribbean island, and one for Mercosur based in Brazil.

the implementation of the existing legislation, assisting with the organization of water management systems at the river basin level, transferring hydraulic infrastructure systems to users, participating in river and lake restoration programmes, recovering natural water courses in urban areas, developing plans for international river basin management, and advising governments on the privatization of water-related utilities. Consultancy requests have increased exponentially in recent years. Central governments are no longer the only ones to request support. In the wake of decentralization and the privatization of water management and use, the demand for consultancy now comes from public and private organizations at national, state, provincial, departmental, and local levels, as well as universities, multilateral and bilateral aid agencies, non-governmental organizations, water-user commissions, public utilities, banks, and river basin organizations.

This growing demand reveals a need for principles, processes, and practices to enable the different actors to proceed correctly in multiple water-use management and public utility regulation. As a general rule, the countries of the region lack these basic instruments by which to co-ordinate the work on a large scale, covering many regions and situations simultaneously and making the most of the scarce resources available.

In spite of the lack of support, the need to improve the management of water resources is acquiring prominence on Latin American and Caribbean government agendas, not only nationally but also at the state and local levels. The various effects of greater private sector participation in water-related public utility companies and in the need for the management of the hydraulic structures as well as the irrigation systems, together with the decentralization of environmental management functions to the municipal level, including the management of river basins and streams, has generated increasing demand for information and technical assistance on this subject. This demand has been further boosted by the creation of numerous but weak river basin organizations, or plans to do so, by debates on water law reform, and by the disastrous effects of flooding and increased water pollution. Even though from the hydrological viewpoint, water resources should be managed in accordance with the concept of the river basin, it is not so easy to accomplish such goals when confronted with the political jurisdiction and administrative organization prevailing in each country. There are invariably problems in implementing such an approach because most of the countries of the region have a long-standing tradition of centralized public administration. Attempts to apply the concept of water management at the river basin level in these countries have generally been only partially successful.

This revival of interest in the river basin as the most appropriate unit for water management is due mainly to the fact that it is precisely at this level that it appears more feasible to achieve a better integration between all parties, whether public or private, and whether their concern with water management is for purposes of production or conservation. Furthermore, water resource management at the river basin level is increasingly considered to be the most appropriate way of absorbing the environmental costs of use of water resources. Nevertheless, there is still a strong emphasis on the physical components of the systems or on activities and investments in the sector, while the organizational component for the establishment of river basin entities, which undoubtedly constitutes the most important aspect of this approach to water resource management, has scarcely been developed.

Different Approaches and Definitions for Water Resources Management at River Basin Level

Management of water at river basin level is not new in the region, but despite this there is still no consensus on definitions that spell out the objectives of that management. The lack of conceptual clarity on the subject impairs the exchange of ideas and experiences, particularly between professionals of different countries, causes overlapping of functions, and hinders the formulation of policies and laws on the subject.

Inconsistencies in the use and meaning of many of the terms relating to river basin management suggest the convenience of defining and classifying such concepts. Table 7.1 summarizes and classifies concepts related to river basin management in Latin America and the Caribbean. A matrix format is used to set out the stages of the management process as they relate to the objectives defined by the elements and resources covered by the management. This layout has been chosen to enhance understanding of the actions that may be co-ordinated in a river basin, and the purpose of such co-ordination. Moreover, it was considered useful to clarify other complexities arising from differences in terminology between English and Spanish. Hence the decision to include entries in both languages; it is hoped that understanding of the Spanish term will be enhanced by a comparison with the original concept.

Table 7.1 sets out two groups of factors, indicating the terminology used for each case:

I. The stages in a river basin management process (1, 2, and 3):

- Preliminary stage (1): studies, formulation of plans, and projects.
- Intermediate stage (2): the investment stage for river basin development with a view to the use and management of its natural resources for purposes of economic and social development. This stage corresponds to the notion of 'development' as in 'river basin development', 'water resources development' (the corresponding term in Spanish being 'desarrollo de cuencas' or 'desarrollo de recursos hídricos' or 'desarrollo de recursos hidráulicos').
- Permanent stage (3): the operation and maintenance stage of structures and management and conservation of natural resources and elements. This phase corresponds to the notion of 'management' (a term which has as many as four meanings in Spanish: 'gestión', 'administración', 'ordenamiento', and 'manejo'). In general, 'water resources

Table 7.1: Management at the River Basin Level: Stages and Objectives

	River Basin Management Objectives		
Management Stages	Integrated Use and Management	Use and Management of All Natural Resources	Water Resources Management (Integrated or Sectoral)
	(a)	(b)	(c)
(1) Preliminary stage		Studies, plans, and projects ('ordenamiento de cuencas')	
(2) Intermediate stage (investment)	River basin development ('desarrollo integrado de cuencas' or 'desarrollo regional')	Natural resources development ('desarrollo' or 'aprovechamiento de recursos naturales')	Water resources development ('desarrollo' or 'aprovechamiento de recursos hídricos')
(3) Permanent stage (operation, maintenance, management, and conservation)	Environmental management ('gestión ambiental')	Natural resources management ('gestión' or 'manejo de recursos naturales') 'Watershed management' ('manejo' or 'ordenación de cuencas')	Water resources management ('gestión' or 'administración del agua')

Source: Dourojeanni, A., 1994, *Políticas públicas para el desarrollo sustentable: la gestión integrada de cuencas,* Economic Commission for Latin America and the Caribbean (ECLAC) and Centro Interamericano de Desarrollo e Investigación Ambiental y Territorial (CIDIAT) and ECLAC, 1994, *Políticas públicas para el desarrollo sustentable: la gestión integrada de cuencas,* LC/R.1399, 21 June 1994, Santiago, Chile.

management', is translated as 'administración de recursos hídricos', and 'watershed management' as 'manejo de cuencas'). It should be noted that Spanish does not normally make a distinction between the concepts 'watershed' and 'river basin', both of these being translated as 'cuencas hidrográficas', although some effort has been made to differentiate between the two by using terms such as 'cuenca fluvial' and 'hoya hidrográfica' to refer to 'river basin', and 'cuenca de alta montaña' or 'cuenca de captación' to render the idea of 'watershed'.

II. Natural resources and elements that are considered in the process of river basin management (letters a, b, and c):

- Group (a): all the elements, resources, and infrastructure for development of a river basin.
- Group (b): all the natural elements and resources to be found in a river basin.
- Group (c): only water resources.

This form of terminological analysis is unique and it is hoped that it may be helpful in classifying concepts of the objectives of river basin management. Table 7.1 shows clearly that the most complete type of management at the river basin level is indicated in column (a), under the heading 'river basin development' at the intermediate stage and 'environmental management' in the permanent phase. This approach amounts to applying regional development and environmental management techniques at the river basin level. It is an approach that gained currency in Latin America following the success of the Tennessee Valley Authority in the United States, an approach that attracted followers in Mexico, Colombia, Brazil, and Peru. Agencies responsible for this type of management are usually referred to as water corporations or commissions. Most originated and developed out of major investment projects.

The intermediate river basin management level is shown under column (b) and includes activities aimed at the co-ordination of the development ('natural resources development') and management of all the natural resources to be found in a river basin ('natural resources management'). This level of systematic management of all natural resources in a river basin (management of the use of a river basin according to its capacity and purpose) has not been applied comprehensively in the region.

There are no systems or entities that facilitate co-ordination of the activities of use and management of the natural resources in a river basin. However, there are many watershed management programmes and projects ('manejo de cuencas'). Watershed management has become a sub-item or part of this integrated approach to natural elements and natural resources management.

The traditional approach to watershed management aimed at regulating the runoff of water (a concept that originated and was first applied in the United States) is part of the approach to natural resources management. Watershed management is therefore a mixed activity, linked to management and conservation of all natural elements and resources as well as water management itself.

The third level of management, which is shown in column (c), is geared towards co-ordination of investments in water resources development and subsequent management thereof. It is the best known level of river basin management in the countries of the region and it is at this level that most of the studies and investments in hydroelectricity, irrigation and drainage, drinking water, and flood control are conducted.

In Latin America and the Caribbean, it is normal for the intermediate phase ('development') geared towards planning and execution of investment projects, in particular hydroelectric projects, to be governed by strong systems of management. This is largely due to the fact that it is a phase that normally benefits from substantial financial backing, political support, and interest on the part of the banking sector.

Conversely, the permanent phase ('management'), involving the day-to-day management or administration (for example, of water, use of flood-prone areas, pollution control, or use of hillsides and operation and maintenance of waterworks except in the hydroelectric sector and some drinking water services) was generally very poor. This is the phase that needs to be improved in all fronts.

Development of Integrated River Basin Management

The development of river basin management in the countries of the region has been neither uniform nor stable. Management systems have been changing erratically, giving rise to many cases where in the past, management, at least of water resources, has tended to be more integrated than at present.

In its initial stages, co-ordination of activities at the river basin level was limited. Work was done at this level in order to solve problems as they arose and satisfy specific or sectorial demands for water, supplying water for population centres or irrigation, controlling floods, and building hydroelectric power stations.

The next step was to operate and maintain the structures thus constructed. This management was limited to the existing structures without any particular interest in multiple use of water resources or in managing the river basin area, that is to say the natural resources of the river basin. Thus, a series of water management systems were implemented in the region, many of which were developed solely for sectorial water use.

In the late 1940s, corporations were set up for the integrated development of river basins, that is, for regional development at the river basin level. Starting from the construction of water projects these corporations were set out to embrace extensive areas under their jurisdiction and to invest in a number of sectors.

In the 1970s emerged the concept of 'watershed management' mainly with the aim of reducing silting up in dams and to control landslides or floods. There are very few instances in which all the natural resources of the river basin are managed. Integrated agricultural, forestry, and livestock projects have helped to improve this aspect but do not compensate for the lack of a well-co-ordinated system for the management of the natural resources of river basins.

The environmental dimension began to be taken into account in Latin America only at the end of the 1970s, that is, some five to seven years after the United Nations Conference on the Human Environment was held in Stockholm in 1972. First came environmental impact studies, and later environmental quality analyses. To a large extent, environmental management at the river basin level did not go beyond the phase of studies and proposals for forming organizations.

A look at Table 7.1 is necessary to understand this development and to identify the different steps in management that cover the entire river basin depending on the phase of execution and the resources to be managed. Table 7.1 shows a total of seven steps (intermediate and permanent phases) for river basin management: three geared to river basin development and four to the control, administration, or management of the environment, natural resources, or water resources.

The chronological order followed in Latin America in co-ordinating actions at the river basin level is as follows:

- First, the question of water control and use in river basins is approached through the construction of water projects ('water resources development').
- Second, the question of the management of water in river basins is tackled ('water resources management').
- Third, there is then a direct transition to 'river basin development'.
- Fourth, the question of 'watershed management' is taken up, especially with a view to controlling the erosion that affects existing dams and preventing landslides and mudslides.
- Fifth, there is then a direct transition to a consideration of the issue of 'environmental management'.

What stands out in this evolution is that there has been an abrupt decision to co-ordinate, at least on paper, environmental management at the river basin and regional levels, without yet having fully co-ordinated the measures for the development and management of all natural resources of a river basin. It will be remembered, however, that if natural resources are not managed in a co-ordinated manner, not even water, then it will be impossible to undertake environmental management. The first step should then be to manage the water resources in an integrated manner and then the other natural resources associated with them.

A review of the history of river basin organizations shows that many never become more than, at best, 'action-coordination systems', and they somehow managed to get some studies on river basin carried out. Historically, some river basin organizations were even created for the specific purpose of sponsoring a study or plan, often carried out by groups of consultants hired temporarily for the purpose. In other words, many short-lived 'river basin organizations' were only intended to direct the execution of inventories, studies, assessments, or diagnoses, or draw up river basin development plans that were somewhat more complete than usual. Many of the studies on individual river basins that are currently available have been conducted by institutes of natural resources or by government ministries; these tend to produce the same results as the integrated river basin studies conducted by temporary river basin agencies.

In other cases, river basin organizations are, in practice, the management structures of investment projects corresponding to major water works in the river basin. The names given to these organizations also tend to be varied, the most common being corporations, commissions, or agencies, or simply 'programmes' or 'special projects' that have been responsible for executing water development investment projects in one or more river basins. Likewise, there have also been many national-level projects devoted to a single type of activity, which have been responsible for simultaneous studies in many river basins. These are what are known as 'national programmes' such as those targeting flood control, watercourse stabilization, soil conservation, drainage and land reclamation, river basin management, or rural electrification, to mention a few examples. Some of these projects have been co-ordinated at river basin level, but most of the national programmes have been run independently.

With the increasing drive for municipal participation in environmental management and the acknowledgement of the vital importance of broad public participation in river basin management programmes, a new focus has developed on the issue of managing river basins and bodies of water that are shared by urban areas and several municipalities. Local officials have become the most recent 'clients' in need of working methodologies on river basin management and recovery of watercourses, with the participation of the inhabitants of their administrative areas.

A variety of historical circumstances have brought about substantial progress in establishing and operating river basin organizations, such as the 25 river basin councils recently created in Mexico. There are also some river basin organizations that have been in operation for several decades. In general, in the course of their existence, they have undergone several changes of name, responsibility, or degree of autonomy. None of them, however, are guaranteed to survive unless they adapt to the changing situations in politics, the economy, and demands of the population. Although the efficient operation of a river basin organization does not ensure its continuity, it does give it a certain degree of security in the face of the institutional changes that may occur in any given country.

Prior to proposing the establishment of a new river basin organization in a country it is, therefore, useful to analyse the historical development of similar organizations. It is a worthwhile exercise to look for explanations of why some of these bodies continue to exist years after their creation, while others have disappeared.

Processes Involved in River Basin Management

Setting up any kind of river basin organization, with a view to river basin management under any of its modalities, entails a series of ongoing processes that can be implemented in parallel. The processes that are particularly worthy of further analysis are: (i) communication and awareness-raising; (ii) formation of alliances and agreements; (iii) legalization of operations; (iv) scenario development, evaluation, and diagnosis; (v) operational consolidation of each water user; (vi) administrative organization; (vii) economic valuation and preparation of strategies; (viii) operation of the shared hydraulic system; (ix) conservation of water bodies, natural habitats, and biodiversity; and (x) pollution control, stream corridor restoration, and recovery of rural and urban drainage capacity. These processes can be divided into three groups: a central co-ordination process, a group of socio-economic processes, and a group of physical and technical processes.

Communication and awareness-raising. Awareness-raising campaigns through whatever media are available are to be recommended before proposing the establishment of any river basin organization. It is a good idea to explain to the actors involved in managing the water resources of a river basin why an agency to co-ordinate their efforts is useful and necessary. This stage also serves to gather information, identify conflicts, and compile literature. It is worthwhile to establish which bodies or organizations are operating in the basin, which of them distribute the water, how they measure distribution, if they keep water quality records, if they have emergency programmes, and, in general, how they operate the existing water systems and with what resources.

Formation of alliances and agreements. The actors involved should set up a preliminary alliance to take action that will gradually progress toward the establishment of an overall system of river basin management. The scope of the alliance can be widened as time goes on but, initially, it is usually easier if the actors set a specific objective for their action (clean-up of a lake or river, reforesting a river bank, administering the water of a river, or canal used by several users, managing the banks and course of a river or any other subject that is of interest to more than one actor). The actors may include public or private groups, non-governmental organizations, municipalities, universities, and professional organizations. Alliances must be formally established and set concrete goals for their work. Ultimately, this activity is expected to give rise to roundtables for co-ordination and dialogue. The list of actors who are invited to take part must be flexible, since it will vary from one situation to another.

Legalization of operations. The legal framework for a river basin organization can be consolidated gradually. If there is no specific legislation under which to create a river basin management system, the parties could start with a simple agreement to carry out a project. The final objective of the process, however, is to give the river basin management system legal personality and clearly identifiable competencies to manage the water in the basin (collection of charges, monitoring, etc.), either directly or by co-ordinating the actions of responsible organizations. There are several ways of affording legal status to actions relating to river basin management, including ministerial resolutions establishing special programmes and projects and responsibilities which are assigned by law to municipalities, ministries, or institutes, which then give their actions legal status through the modalities of ordinances, regulations, and other directives.

Scenario development, evaluation, and diagnosis. Once a minimum degree of commitment and agreement has been obtained among the actors in the alliance about what they want to achieve in the river basin through their co-ordinated action, the existing situation must be evaluated in order to arrive at a diagnosis. This will require the participation of an interdisciplinary team and can be defined as a management procedure for sustainable development. The actors must be encouraged to participate in a public debate about the issues to be addressed. It is also important to promote the use of geographical information systems and, in general, of all available techniques for describing what is happening in the basin, who the affected and responsible parties are, and what are the costs and benefits involved in the programme of action.

Operational consolidation of each water user. The aim of this process is to help each actor involved in managing the water and the river basin to ensure that they are complying fully with their responsibilities. For example, support should be given to organizations of agricultural users, drinking water and sanitation services, mining, fisheries and recreational users, and, in general, all those actors who in some way alter the flow of water in the basin, to ensure that their practices conform to the highest standards possible. This consolidation process includes providing support to local governments to help them comply with their environmental responsibilities and to ministries—such as health ministries—to help them discharge their role of environmental quality control, and to other entities including non-governmental organizations.

Administrative organization. All the stages must be carried out within an adequate administrative framework, including the collection of charges, registering of actors, accounting, financial controls, monitoring and ensuring compliance, procurement of equipment, and hiring of staff and consultants. The administrative system will become more complex as the process advances. If the organization is to survive, it must make itself indispensable, and that will only happen if it generates confidence in its financial management and the quality of its work. The professionals who make up the administrative system must be suitably qualified.

Economic valuation and preparation of strategies. Plans are written strategies, and are presented in the form of programmes of work or projects which have due technical and financial backing. Once it has started, the process of planning is never concluded. Planning should be seen as equivalent to building a system of information and rules, standards, and criteria that facilitate decision-making among multiple actors. The factors which are used to calculate costs and benefits, design strategies, and draw up a plan come from the stages of identifying the actors, their criteria, problems, and objectives, building shared scenarios, evaluating the existing situation, making diagnoses, and identifying obstacles and restrictions. The plan should serve to communicate intentions and co-ordinate where necessary.

Operation of the shared hydraulic system. Qualified technicians are needed to operate and maintain the hydraulic system built in the river basin and to support water conservation and management, and the many users in the river basin must also participate in the process. The basin's rivers and hydraulic systems must also be equipped with a series of water monitoring stations and satellite information systems, or these must be reinforced if they already exist. In general, the organization needs to be sufficiently equipped to be able to keep track of situations and plan ahead. Modern communications systems are essential to enable the overall system to function correctly.

Conservation of water bodies, natural habitats, and biodiversity. It is not enough to merely operate the hydraulic systems built. An enormous amount of work is required to recover damaged areas along riverbanks and riversides and rehabilitate biological habitats. It is essential to mitigate the effects of conflicts related to water and river basin management by ensuring that plans for the use and occupation of the territory respect—as far as possible—the natural catchment and water-flow conditions in the basin. This is necessary to maintain all the river basin's original functions, in particular to conserve biodiversity and the landscape. This process requires town planners to take account of natural watercourses, with normal and seasonal flows, in their decision-making.

Pollution control, stream corridor restoration, and recovery of rural and urban drainage capacity. In most river basins, especially in urban areas, this process entails reversing situations that have already profoundly altered watercourses and flows. This is a long task and likely to be the most challenging of all. It is not possible to conserve basins or watercourses if they have already deteriorated totally. While industrialized countries are in the process of rehabilitating stream corridors, most developing countries are in the process of destroying them.

This analysis is clearly not intended to be complete. Neither does it address how to combine all these processes in a flow diagram of work, incorporating the activities, staff, and time for each action. The implementation of the stages described above will be greatly aided if theoretical and practical data is complied to support the establishment of the river basin organization. This can be complemented with additional information such as an evaluation of the knowledge of water users, the actors who will be involved in managing the water in the river basin, their criteria on multiple water-use management, the problems and conflicts involved in shared management, and the objectives they are pursuing. It is also necessary to carry out a comparative analysis of the past and present experience of attempts to create such organizations within the country, and if possible in more than one country, whether these have been successful or not.

A particularly important point for making the processes of creating and consolidating a river basin organization as smooth as possible is to begin while the hydraulic works are still being built, whether they are State or privately operated. Commonly, the 'master plan' for integrated river basin management is not thought of until the works are finished. Still worse is that this often means that no resources are available for setting up the operative system—which amounts to much more than making a plan—including funding for complementary communication works and monitoring systems. At least 5–10 per cent of the cost of the major hydraulic works should be allocated to establishing the management system—including the necessary infrastructure. No less than ten years should be allowed for consolidation, especially in river basins featuring a combination of formal and informal actors and low-income groups.

The Long Process of Creating the Legal Framework for River Basin Organizations

In order to analyse the institutional structures for river basin management, it is essential to attempt to distinguish between the many variations they adopt. There are three basic types of structures in river basin management organizations:

- *Management structure.* Management structures vary depending on the extent to which the different actors participate in the management process. The name given to the river basin organization does not necessarily reflect their degree of participation in the decision-making process but it does, at least, indicate the original intention. The most common formulas are 'river basin commissions', 'river basin committees', 'river basin councils', and 'river basin agencies', which display a wide range of types of participation by the actors involved in the decision-making process. In other cases, the management structure consists of a board of directors, which may be composed of government officials only or may include users, non-governmental organizations, universities, etc. The board of directors must have the power to decide, resolve, and enforce agreements (it should not be merely an advisory or co-ordinating body).

- *Operational structure.* An operational structure is the body which puts the decisions of the management group into practice. It executes actions and processes, either directly or through consultants and contractors. The operational structure of a river basin organization must have highly qualified personnel. They are the 'agency' in the strict sense, although they may be known by other titles, such as executive office, technical group, technical office, corporation, or even institute, for example. The operational structure is the one responsible for providing the studies and information that the management group needs to take decisions.
- *Financial structure.* The body responsible for raising financial resources is one of the most difficult to design. In the countries of the region it is common to find that financial resources for river basin management are only available at the phase of executing hydraulic works, which is obviously not the solution for a river basin organization that is intended to be permanent. Few 'models' of financial structure are transferable from one country to another. The polluter pays principle, aids and incentives are a good option but are clearly insufficient and even inapplicable to many of the region's river basins which are characterized by informal settlements and producers. Any financing proposal must

take into account the situation of the country, region, and river basin.

The creation of formal institutional conditions for river basin organizations is at varying stages of progress in the region. Without a doubt, the best scenario involves national legislation serving as a regulatory framework for the process of creating river basin organizations while also providing for the possibility of adopting alternative approaches at the state, provincial, and regional levels, in line with the country's political and administrative structure.

In federal countries, and countries with regions which have greater or lesser degrees of autonomy, the legal framework for establishing river basin organizations is usually established at the respective administrative level (state, province, or region). In some cases, the framework is jointly created by mayors who share a river basin or by simple agreements between the main users of water in these areas.

The legal provisions establishing these organizations should usually be part of a wider framework of water legislation, as occurs in Mexico and Brazil. However, there are also situations in which the national level legal provisions come under legislation on decentralization, environmental laws which include provisions on territorial organization, investment promotion legislation, laws on national investment programmes or projects, or other variations arising from proposals by different ministries or regional governments.

Transboundary river basin development agreements can also provide a basis for formalizing the framework of river basin management. These agreements tend to be lasting, indeed much more so than agreements concluded at the national level. Some of the river basin organizations which have survived longest—albeit with some changes of name and responsibilities—are precisely those which come under international treaties involving bilateral or multilateral commitments.

Another major catalyst—and technical factor—in stabilizing and conferring legal status on river basin organizations are bilateral technical assistance agreements. These agreements have the virtue of providing a legal framework for the creation and operation of river basin organizations through agreements with banks or with partner countries. This arrangement puts the respective organization in a better position to withstand at least one direct attack, which can come in the form of a change in management, a change in attitude by some official, or the sudden structural and operational transformation of the public agency responsible for controlling it.

The process of providing a legal framework for any type of river basin organization is slow and many fall along the wayside. The fact that a law is passed to establish such an organization represents no guarantee whatsoever that it will be implemented. The approval of legislation is only a preparatory step, which must be made in parallel with many other actions, particularly in relation to organizing and implementing the formulas needed to create and operate these organizations. For river basin organizations to become consolidated, they must also be given the capacity to raise their own funding.

Difficulties in Establishing and Operating River Basin Organizations in Latin America

The establishment of river basin organizations very often faces opposition from some of the main users, sometimes from interinstitutional rivals and sometimes because they have to confront or compete openly with regional authorities. Many organizations which have been in operation for years continue to face the same set of conflicts and opposition. Many river basin organizations have succumbed to this problem, as the statistics of some countries show. The organizations which last the longest are those which can rely on their own fund-raising system.

Probably some of the greatest obstacles to the establishment and successful operation of river basin authorities are lack of awareness on the part of public and users of the economic advantages of having such organizations; lack of clarity about their role, which generates potential competition with other authorities and with the public and private sectors; an unrepresentative management level (council, committee, or board); problems with the means and legal basis of raising funds; and the fact that water management at river basin level is often dominated by a sector which has no interest in forming part of a system of shared management.

To establish a river basin organization it is, therefore, necessary to run several processes in parallel. It is strategically advisable to start by acknowledging any type of water administration that already exists in the river basin—whether this is a single sector user, such as irrigation, hydroelectricity, or drinking water supply and sanitation, or various sectors—and involve them in the process right from the beginning. Many past failures or delays in creating river basin organizations are attributable to the neglect of something as fundamental as this.

It is obviously essential to have agreements in place among the public institutions which are involved in water management. Conflicts between State agencies are very injurious to the process and often occur between ministries and agencies, even from the same sector, to the extent that one party may even boycott the initiative. Conflicts sometimes arise between local authorities or provinces and the central government for political reasons, especially if the mayor or governor belongs to the ruling party's opposition.

In general, most of the financial agents of major hydraulic works are guilty of a glaring lack of provision for financing the establishment of river basin organizations to operate and maintain the hydraulic works once they are built. This is usually considered to be allocable to current expenditure of the fiscal budget and not to project expenditure.

Conflicts over the creation of river basin organizations also arise because of the effects of existing legislation, or the lack of it. Sometimes an existing law that provides for the creation of a river basin organization is not flexible enough to allow it to achieve its purpose: it may establish conditions for the participation of actors, composition of boards, or charges which are impracticable. In other cases, there is no legislation on which to base the creation of a river basin organization, afford a legal framework, or provide financial support.

From the hydrological viewpoint, river basins are ideal territorial entities for water resources planning and management. However, in situations where political–administrative jurisdictions do not coincide with the physical boundaries of river basins, many of the decisions that affect the hydrological cycle, water use, and the inhabitants of the basin fail to take into consideration this integrated system as a whole. Furthermore, water resources management is normally fragmented along the lines of user groups, sectors entrusted with overseeing the resource, types of use, the source of catchment, and other similarly arbitrary criteria. An integrated system and a shared resource are administered in a piecemeal fashion, and as a result more situations of conflict over water resources management occur when they should be avoided, minimized, or resolved. *The challenge we face, therefore, is to create competencies for governability over areas delimited by natural factors, such as river basins, which do not correspond to traditional forms of government over political–administrative boundaries, such as states, provinces, regions, and municipalities.*

How Can the Failure of Water Resources and River Basin Management Processes be Prevented?

Attempts to establish water resources and river basin management systems usually fail because proposals for the creation of the pertinent organizations, whether in the form of authorities, agencies, or any other body, are presented in a relatively superficial manner. Generally the aim is to give systems a holistic focus. Hence they should: (i) be economically efficient, self-sustaining, and competitive; (ii) have a social orientation, promote social equity, and be environmentally responsible; and (iii) involve both public and private sectors, promote civic participation, and take a conciliatory rather than an authoritarian approach. In essence, the objective is to create a superior body responsible for fostering sustainable development.

Experience shows that the creation of any organization that performs at least some of the basic functions, such as preventing, reducing, or solving disputes among water users, should be a gradual process. The initial step should be to gather information on public policies in regard to water resources and the economy; the features of water resources and river basin management; the characteristics of water management systems and the actors involved; and the most appropriate methods of operation for public or private organizations responsible for managing water and natural resources in a river basin.

Viewed from this perspective, it may be very useful to analyse policy declarations in terms of a methodological sequence which seeks to direct management procedures towards sustainable development. It is suggested that in order to execute actions, it is necessary to:

   (i) identify the actors involved in the management process;
   (ii) analyse the actors' criteria (policies, principles, etc.);
   (iii) identify any problems related to these criteria;
   (iv) identify what the actors' objectives are;
   (v) define the spheres within which it is hoped to attain these objectives;
   (vi) identify constraints on the attainment of these objectives;
   (vii) propose solutions for overcoming these constraints;
  (viii) decide on the strategies to be applied in order to achieve solutions;
   (ix) design programmes and projects for carrying out the selected strategies and evaluate them; and
   (x) execute both one-off and ongoing programmes and projects.

In accordance with this sequence, policy formulation takes place mainly at the stage when criteria for action and the actors' objectives need to be specified. These criteria are for the most part declarations of intent. By contrast, policies for executing actions can only be formulated once the solutions and strategies have been designed. Thus, water policy formulation needs to be undertaken step by step, in a systematic way, so as not to overlook aspects critical to successful implementation.

Water policy formulation in the countries of the region has seldom been carried out in a rigorous way. Generally speaking, policy formulation is ad hoc, and does not follow any established procedure. Water policies in the region have at various times emphasized the preparation of plans, the formulation of laws, the creation of new entities, and so on. However, it is a matter of concern that the vast majority of these proposals are not properly harmonized. The measures taken in this context are piecemeal, their objectives limited to, for example, avoiding inconsistency with an economic system, reinforcing other laws, mitigating specific conflicts that arise from time to time among users, satisfying the demand of certain groups of voters or facilitating a particular decentralization project. In such circumstances, the water policies formulated are normally incomplete. For example, decentralization in some countries has led to profound contradictions between development policies and water policies, with the result that river basin organizations attached to the central government sometimes find themselves subordinate to two or even three authorities, because the river basin under their control has been divided by regional boundaries.

Water policies should fit neatly with national development policies, but it should also be pointed out that both water resources and processes to develop them have certain features which, if neglected, give rise to huge contradictions. The unique features of water as an economic resource demand, if not a dominant role for the State, at least joint management by the State and users of supply at the river basin or interconnected system level. This is the only way to resolve any conflicts that may arise, to make resources available to deal with shared problems and to control externalities, natural monopolies, and other aspects that require regulation.

As the consequences of water management policies in force are often unknown, it is difficult to come up with a way to improve them. In other words, if there is a lack of information about how water development policies are currently working (causes and effects), it is hard to decide what to do to make them more effective. Many countries do not maintain an up-to-date register of laws dealing with water resources and watershed management. Countries also sometimes lack a register of users of river basin or water systems, as well as an inventory of studies on each system or of investments made in water infrastructure works in each basin. It is not known to what extent policy declarations and official rulings on functions are implemented in practice. A large number of government agencies do not have sufficient resources to perform the tasks they are set. Until now, most water policies that stem from changes in economic policy remain little more than declarations or policies of intent. In many cases, without any deeper analysis, policies of intent have become laws of intent, and this has generated serious gaps, especially in terms of instruments to implement the laws. In several cases, the spirit of the policy bears little relationship to the provisions of the law or to the results it achieves.

References

CEPAL, 'La pequeña cuenca de montaña en la gestión del desarrollo y en la conservación de los recursos naturales', LC/R.626, 21 December 1987, Santiago, Chile.

———, '¿Qué hacer después de Rio?: lo que no se hizo antes de Estocolmo', LC/R.1229, 19 November 1993, Santiago, Chile.

———, 'Políticas de gestión integral del agua y políticas económicas", LC/L.781, 5 November 1993, Santiago, Chile.

———, 'Políticas públicas para el desarrollo sustentable: la gestión integrada de cuencas', LC/R.1399, 21 June 1994, Santiago, Chile.

———, 'Procedimientos de gestión para el desarrollo sustentable (un breve glosario)', LC/R.1450, 20 September 1994, Santiago, Chile.

———, 'Economía y ecología: dos ciencias y una responsabilidad frente a la naturaleza', LC/R.1457, 4 October 1994, Santiago, Chile.

————, 'Los procesos naturales y artificiales en la transformación de la estructura productiva', LC/R.1459, 6 October 1994, Santiago, Chile.

————, 'Planes y marcos regulatorios para la gestión integrada de cuencas', LC/R.1487, 23 January 1995, Santiago, Chile.

————, 'Conceptualización, modelaje y operacionalización del desarrollo sustentable ¿Tarea factible?', LC/R.1620, 22 January 1996, Santiago, Chile.

————, 'Reflexiones sobre estrategias territoriales para el desarrollo sostenible', LC/G.1944, 29 November 1996, Santiago, Chile.

————, 'Creación de entidades de cuenca en América Latina y el Caribe', LC/R.1739, 10 July de 1997, Santiago, Chile.

————, 'Informe del II Taller de Gerentes de Organismos de Cuenca en América Latina y el Caribe (Santiago de Chile, 11-13 December 1997)', LC/R.1802, 12 February 1998, Santiago, Chile.

————, 'Reflections on territorial strategies for sustainable development', (LC/G.1944, 18 March 1998), Santiago, Chile.

————, 'Informe del III Taller de Gerentes de Organismos de Cuenca en América Latina y el Caribe (Buenos Aires, Argentina, 16-18 November 1998)', LC/R.1926, 3 August 1999, Santiago, Chile.

————, 'Report on the Second Workshop for Managers of River Basin Authorities in Latin America and the Caribbean. Santiago, Chile, 11-13 December 1997', LC/R.1802, 1 September 1999, Santiago, Chile.

Dourojeanni, A., 1997, 'Management procedures for sustainable development (applicable to municipalities, micro-regions and river basins', LC/L.1053, Serie Medio Ambiente y Desarrollo No. 3, Santiago, Chile.

————, 1999, 'La dinámica del desarrollo sustentable y sostenible', LC/R.1925, Santiago, Chile.

————, 2000, Procedimientos de gestión para el desarrollo sustentable' LC/L.1413-P, Serie Manuales No. 10, Santiago, Chile.

————, 2001, 'Water management at the river basin level: challenges in Latin America', Serie Recursos Naturales e Infraestructura No. 29, Santiago, Chile.

Dourojeanni, A. and A. Jouravlev, 1999, 'Gestión de cuencas y ríos vinculados con centros urbanos', LC/R.1948, Santiago, Chile.

Röling, N., Gateway to the Global Garden: beta/gamma science for dealing with ecological rationality, Eight Annual Hopper Lecture, 24 October 2000, University of Guelph, Canada.

Index